I0482668

NIST GCR 09-921

Analysis of FDS Thermal Detector Response Prediction Capability

Prepared for
U.S. Department of Commerce
Building and Fire Research Laboratory
National Institute of Standards and Technology
Gaithersburg, MD 20899-8664

By
Morgan J. Hurley
Alex Munguia
Society of Fire Protection Engineers
7315 Wisconsin Ave., #620E
Bethesda, MD 20814

February 2007
Issued January 2009

U.S. Department of Commerce
Carlos M. Gutierrez, Secretary

National Institute of Standards and Technology
Patrick D. Gallagher, Deputy Director

Notice

This report was prepared for the Building and Fire Research Laboratory of the National Institute of Standards and Technology under Grant number 60NANB5D1207. The statement and conclusions contained in this report are those of the authors and do not necessarily reflect the views of the National Institute of Standards and Technology or the Building and Fire Research Laboratory.

Analysis of FDS Thermal Detector Response Prediction Capability

Prepared for
National Institute of Standards and Technology
Building and Fire Research Laboratory
Gaithersburg, MD 20899

Grant #60NANB5D1207

Morgan J. Hurley, P.E.
Alex Munguia, EIT[*]

Society of Fire Protection Engineers
7315 Wisconsin Ave., #620E
Bethesda, MD 20814

February 28, 2007

[*] Presently with Schirmer Engineering

ABSTRACT

Predictions of fire plume and ceiling jet temperature and the response of thermal detectors from NIST's Fire Dynamics Simulator (FDS) were compared to data from a series of full-scale tests conducted by Underwriters Laboratory. The tests were conducted in a 36.6 m × 36.6 m compartment with ceiling heights ranging from 3.0 meters to 12.2 meters. Heat release rates followed a modified t^2 growth profile. Thermocouples attached to brass disks were used to simulate thermal detectors.

FDS simulations were conducted with a grid spacing of 100 mm. A convergence study found that grid-size convergence was achieved outside of the plume region. However, grid convergence was not achieved in the plume region at this grid spacing. Outside of the plume region, FDS predictions were within a factor of 1.9 of test data.

KEY WORDS: FDS; DETACT –QS; model evaluation; fire experiments; heat detectors

INTRODUCTION

In 2002, the Society of Fire Protection Engineers (SFPE) published the *Engineering Guide: Evaluation of the Computer Fire Model DETACT-QS.*[1] This guide was the first comprehensive, independently conducted evaluation of a computer fire model ever published.

DETACT-QS[2] is a fire model that estimates the activation time of thermal detectors and sprinklers. DETACT-QS uses correlations developed by Alpert[3] to predict the temperature and velocity of fire plumes and ceiling jets resulting from a user-defined fire. Thermal detectors and sprinklers are modeled as a lumped mass. DETACT-QS solves an ordinary differential equation using an Euler technique. Despite its age, DETACT-QS is one of the most widely used computer fire models.

Three series of test data were used to evaluate DETACT-QS:

1. A series of tests conducted at Underwriter's Laboratories with an "unconfined" ceiling and ceiling heights ranging from 3 meters to 12 meters that used a heptane spray burner as the fire source

2. Two tests conducted at Factory Mutual that used wood cribs as a fire source

3. Tests conducted in a residential scale room.

SFPE's analysis showed that DETACT-QS predictions were more accurate under some conditions than others. Specifically, it was found that as the radial distance of the detector from the plume centerline increased, predictions generally improved. Similarly, predictions generally improved as the response time index of thermal detectors increased. Since the scope of the analysis was limited to the evaluation of DETACT-QS, development of better predictive methods was not explored.

More recently, Fire Dynamics Simulator[4] (FDS) has been developed. FDS is a computational fluid dynamics model that permits the discretization of a space into user-defined numbers of grid cells. FDS has quickly become a widely used tool in the fire protection engineering community. FDS models a variety of fire phenomena, including prediction of thermal detector response.

However, unlike many other computer fire models in existence, FDS uses a much different technique to model thermal detector response than does DETACT-QS. Like DETACT-QS, FDS uses a lumped-mass model of thermal detectors and a numerical technique to determine the thermal response to local gas temperature and velocity. However, FDS determines the temperature and velocity of fire plumes and ceiling jets using a large eddy simulation technique. Therefore, the conclusions found in the *Engineering Guide: Evaluation of the Computer Fire Model DETACT-QS* are not applicable to FDS.

To determine the capability of FDS to predict thermal detector response, FDS predictions were compared to a subset of the data used to evaluate DETACT-QS. Specifically, the data from the experiments conducted at Underwriter's Laboratories were compared to FDS predictions. FDS version 4.0.6 was used to perform simulations.

TEST DESCRIPTION

Tests were conducted in a 36.6 m x 36.6 m facility with a smooth, flat, horizontal ceiling that measured 30.5 m x 30.5 m.[5] The height of the ceiling was adjustable. Ventilation exhaust at a rate of 28 m³/s was provided above the ceiling such that a smoke layer would not form. Tests were conducted with the ceiling positioned at heights of 3.0, 4.6, 6.1, 7.6, 10.7, and 12.2 meters. A minimum of two replicate tests were conducted at each ceiling elevation.

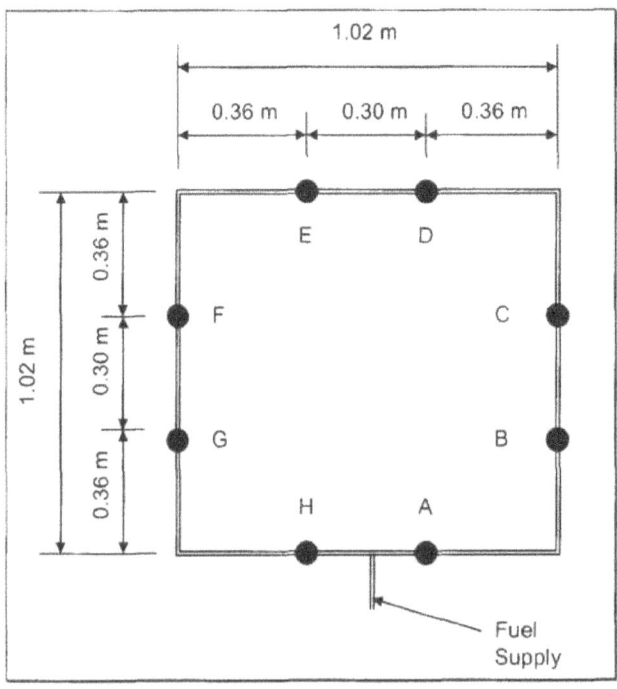

Figure 1 – Burner Configuration

A heptane spray burner was located under the center of the moveable ceiling and elevated 0.6 meters above the floor. The burner was made using 12 mm diameter piping oriented in a square that measured 1.02 meters on a side with two atomizing spray nozzles per side. A diagram of the burner is shown in Figure 1.

Because of the low heptane flow rates used in these tests, only nozzles A, B, D & G were used in the experiments with ceiling heights of 3.0 and 4.6 meters.

The heptane flow rate was controlled manually in an effort to create a growing fire that followed a "medium" growth curve (fire growth coefficient = 0.01172 kW/s².) In the tests with ceiling heights of 3.0 m and 4.6 meters, the heptane flow rate was leveled off to provide constant fire sizes of 1055 kW and 2100 kW, respectively. However, in the other scenarios, the heptane flow rate was increased throughout the duration of the test.

The flow of heptane to the burner was manually controlled using two float type flow meters connected in parallel. The first flow meter had a resolution of 0.08 lpm and a range from 0.68 lpm to 9.1 lpm. The second flow meter had a resolution of 1.1 lpm and a range from 0.91 lpm to 11.4 lpm. Given an approximate density and heat of combustion for heptane of 687 kg/m³ and 44.4 MJ/kg,[6] a theoretical heat release rate can be calculated for the fuel as 30.5 MJ/l. Therefore, the measurable flow rate range of the system was able to provide heat release rates ranging from approximately 350 kW ± 20 kW to 10.4 MW ± 0.3 MW.

Because of the heptane flow instrumentation limitations, it was not possible to precisely follow the desired medium t^2 growth profile. This was particularly true at the early stages of fire growth where the heptane flow necessary to achieve the desired heat release rate was below the resolution of the flow meters. Therefore, adjustments to the medium growth curve were necessary. A time offset of 200 seconds was used to increase the minimum fire size.

The initial fire size varied from experiment to experiment due to the limitations of controlling the heptane flow rate and difficulties experienced in igniting the burner at low flow rates. The burner was ignited by four small pilot fires, which were estimated to have a combined heat release rate between 15 and 20 kW. Also, because the flow measurement occurred remotely from the burner, inaccuracies were introduced by fuel line fill time. The heat release rate achieved from the burner was also affected by incomplete combustion in the heptane spray during the early stages of fire growth. This resulted in the creation of pool fires of varying sizes on the floor during the start of the experiments. Based on these factors and estimations based on the observed fire size, a modification to the medium t^2 growth curve was used to estimate the actual heat release rate achieved.

Table 1 – Estimated Heat Release Rate from Heptane Burner

Time (s)	Heat Release Rate (kW)
0 through 40	$\dot{q} = 0.1875(t + 10)^2$
Time > 40	$\dot{q} = 0.0117(t + 160)^2$

Note to Table 1: The maximum heat release rates in the experiments with 3 m and 4.6 m ceiling heights were 1055 kW and 2100 kW, respectively.

As the fire size increased, the difficulties with accurately measuring the heptane flow were minimized, and it was possible to follow the medium t-squared growth curve more closely. The equations in Table 1 were used to estimate the heat release rate that was achieved from the burner.

The ceiling was constructed of 0.6 m x 1.2 m x 16 mm thick UL fire rated ceiling tiles suspended from 38 mm wide steel angle brackets. Reported[5] thermal properties of the ceiling tiles are provided in Table 2.

Table 2 - Thermal Properties of Ceiling Tiles

Density	313 kg/m^3
Thermal conductivity	0.0611 W/m °C
Specific heat	753 J/kg °C
Thermal diffusivity	2.6 x 10^{-7} m^2/s

Instrumentation consisted of thermocouples to measure temperature. Arrays of thermocouples were provided 100 mm below the ceiling at the plume centerline and at radial distances of 2.2 m, 6.5 m and 10.8 m from the plume centerline. These distances correspond to the radial distances of sprinklers at the corners of squares created by a 3 m x 3m (10 ft x 10 ft) sprinkler spacing with a fire located at the center of the square. See Figure 2.

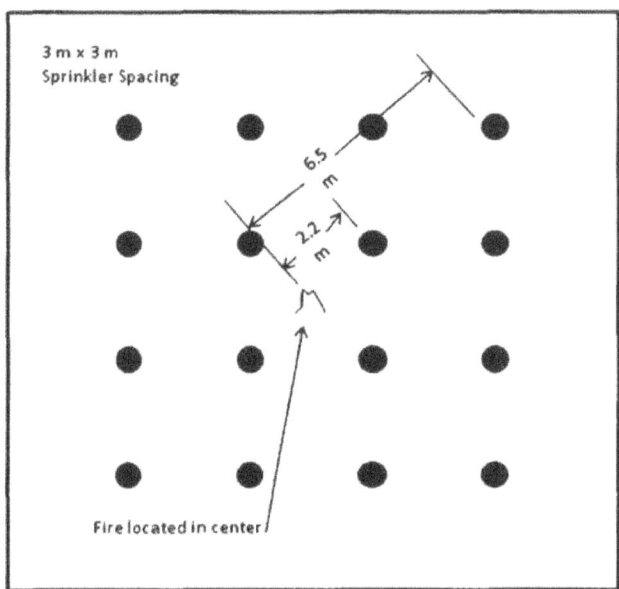

Figure 2 - 3 m x 3 m Sprinkler Spacing

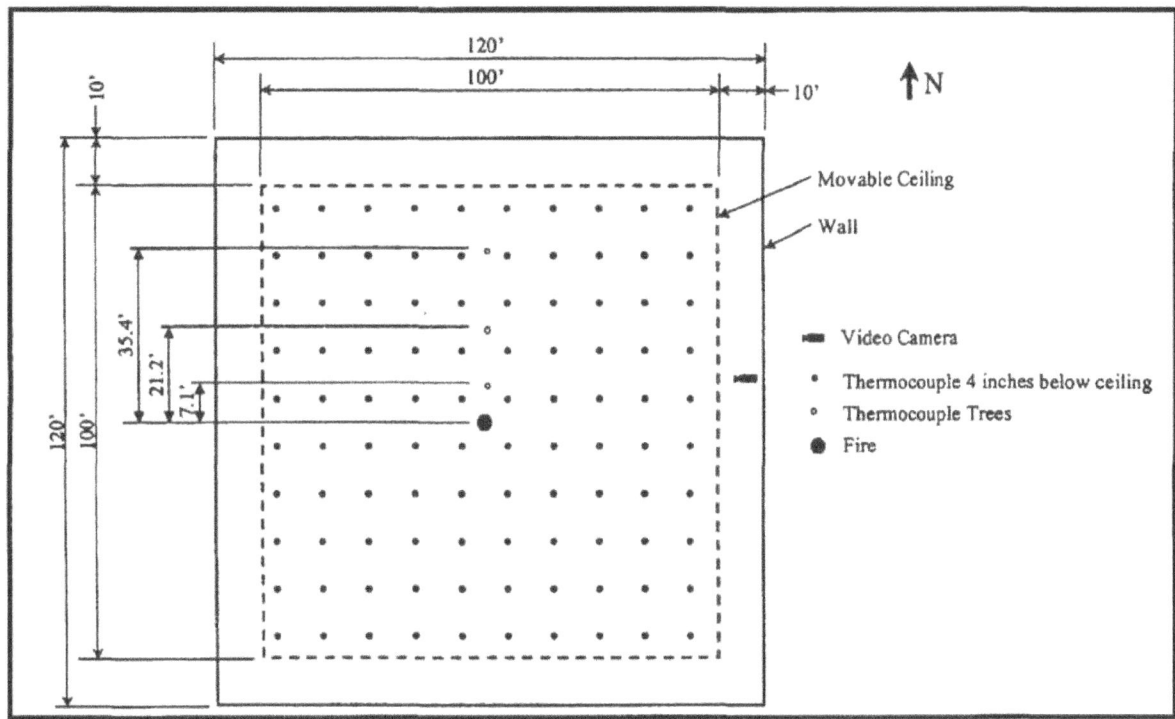

Figure 3 - Reflected Plan View of Experimental Setup

At each thermocouple array, four thermocouples were provided: a type K inconel sheathed thermocouple, and three thermocouples soldered to 25 mm brass disks to simulate heat detectors. The disks were 25.4 mm in diameter and had thicknesses of 0.41 mm, 3.18 mm and 6.54 mm. The thermocouples attached to brass disks were determined to have response time indexes (RTI) of 32 $m^{1/2}$-$s^{1/2}$, 164 $m^{1/2}$-$s^{1/2}$, and 287 $m^{1/2}$-$s^{1/2}$ when tested in accordance with UL 1767.[7] Other thermocouples were used in the testing, but they were not used in the analysis of FDS. Figure 3 shows a reflected ceiling plan of the experimental setup.

MODELING APPROACH

Fire Dynamics Simulator input consists of user prescribed boundary conditions for a user defined computational domain. Model users also specify the grid spacing in each of the Cartesian coordinate directions. Theoretically, as the grid spacing approaches zero, the solutions found should approach the exact solution. However, model run times also increase with the number of grid cells used in a simulation.

Because of the physical size of the experimental facility, it was not possible to run simulations with fine grid resolution within acceptable timeframes. However, it was not necessary to model the entire experimental facility for two reasons: The fire and the instrumentation that was modeled only use a portion of the facility, and (2) because of the physical size of the facility in comparison with the fire size, a significant volume of the space would neither influence, nor be influenced by, the fire.

The remaining space was modeled using a multi-block approach. The first block consisted of a 10 meter by 10 meter computational domain that extended from the floor to the ceiling. The floor was left as the default "inert" surface, and the ceiling was assigned boundary conditions

5

that corresponded to the material properties of the ceiling tiles used in the experimental setup. The four vertical surfaces of the computational domain were opened to the outside of the computational domain. The grid spacing in the two horizontal directions ("X" and "Y") were set as 100 mm. The grid spacing in the vertical direction ("Z") was set to be as close to 100 mm as possible while ensuring that the number of grid cells in the "Z" direction was only divisible by 2, 3, and/or 5.

A second block was used to simulate the ceiling jet area and began where the first block ended. This block also measured 10 meters by 10 meters, but extended from the ceiling to one half of the distance from the floor to the ceiling. A grid spacing of 100 mm was used in the two horizontal directions, and the grid spacing in the vertical direction was identical to that used in the first block. By selecting these grid spacings, the exterior boundaries of the grid cells where the meshes intersected perfectly coincided. Therefore, the transfer of information from one mesh to the other occurs where the meshes intersect.

The top of the second mesh was assigned boundary conditions that corresponded to the ceiling tile used in the experimental setup. The remaining five boundaries of the second mesh were opened to the outside of the computational domain. The volume that was above the moveable ceiling, including the exhaust ventilation, was not modeled.

The reaction was set using the parameters for "heptane" contained in the DATABASE.DATA file that comes with FDS. In the DATABASE.DATA file, the entry for heptane does not designate a radiatiive fraction, so FDS uses the default value of 0.35.

The burner was modeled as an inert box that measured 1 meter by 1 meter by 0.6 meters. This size was 0.02 meters smaller than the burner used in the experiments in each of the horizontal directions. The top of the burner was assigned surface properties that corresponded to the heat release rate of the burner used in the experimental setup. A "ramp" function was used to match the heat release curve provided in Table 1.

Instrumentation arrays were simulated within FDS by placing "thermocouples" and "heat detectors" at locations immediately above the center of the "burner," and radial distances of 2.2, 6.5 and 10.8 meters from the center of the burner measured in the "Y" direction. The instrumentation was placed 0.1 meters (100 mm) below the ceiling. Since heat detectors and thermocouples are not "physical" devices within FDS, at each radial distance the thermocouple and the three heat detectors were located at the same point. Heat detectors were assigned an activation temperature of 1000 °C to ensure that a complete record of device temperatures was recorded by FDS.

Figure 4 illustrates how the space was modeled. The FDS input files used are included in Appendix A.

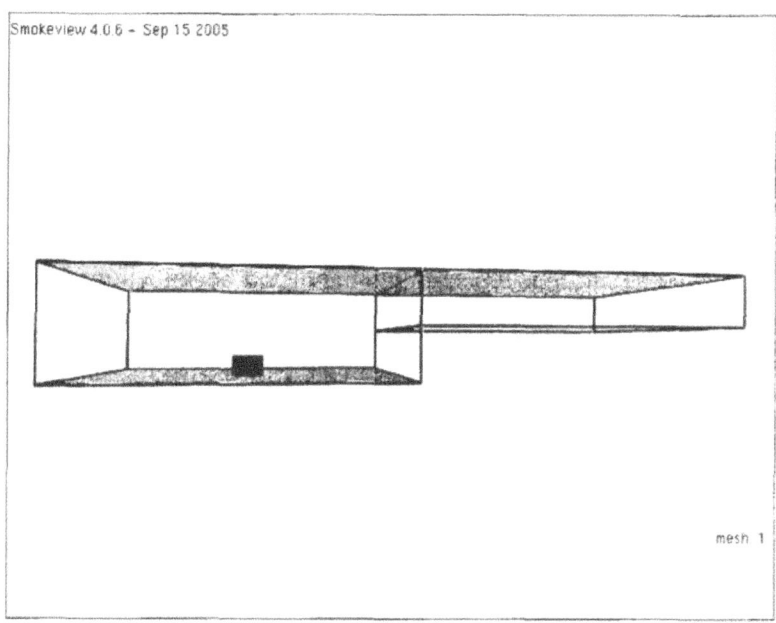

Figure 4 – FDS Representation of Experimental Facility

The ambient temperature was left as the default value of 20 °C. From the test data, it appeared that in some cases the ambient temperature differed from this value by as much as nine degrees. It was not possible to determine the ambient temperature in the tests from the recorded data. First, there was no recording in the test report about the time of ignition. Therefore, the ignition time had to be inferred from the test data at the time when a consistent rise in temperature began. This was also complicated by the fact that small pilot flames were used to ignite the burner, and it was difficult to determine when the temperature rise was caused by the burner and when it was caused by the pilot flames. If the ambient temperature was selected as a time that was clearly before any measurement in temperature rise, there would be an immediate rise in temperature at the time of ignition, which could also affect the results.

A convergence study was conducted by reducing the grid spacing to approximately 66 mm in each of the three Cartesian coordinate directions for the scenarios with 3.0 meter and 6.1 meter ceiling heights.

Thermocouple and heat detector output was imported into a spreadsheet for analysis. FDS "thermocouple" data exhibited a tremendous amount of scatter. To smooth the thermocouple data, the predicted temperature at each time step was averaged with the predicted temperatures during the preceding four time steps and the subsequent four time steps.

EXPERIMENTAL UNCERTAINTY ANALYSIS

Uncertainty associated with the data used in this analysis comes from three sources: (1) uncertainty in thermocouple temperature measurements, (2) uncertainty in fuel flow measurements, and (3) repeatability uncertainty.

Uncertainty in thermocouple measurements is estimated as +/- 2.2 °C based on manufacturer's data.[8]

Which flow meter(s) were used to measure heptane flows was not reported. Since the flow meter with the greater resolution had a range that was 75% of that of the flow meter with the lesser resolution, it was assumed that the flow meter with greater resolution was solely used until the heptane flow had reached the limit of the meter's range. This assumption seems reasonable, in that the meter with the lesser resolution would not be capable of measuring the flow rates that would occur in the early parts of the experiments.

The resolution of the flow meter with the greatest resolution was 0.08 lpm. Because readings on the flow gage could be made to an accuracy of half the resolution, the estimated uncertainty in flow rate would be +/- 0.04 liters per minute. This corresponds to an uncertainty in heat release rate of +/- 20 kW. The uncertainty in heptane flow during the start-up portion of the text is likely greater (and more difficult to quantify.) Therefore, data was not used from before 100 seconds into the experiments. Uncertainty associated with human error in reading the flow meters or in manually controlling the fuel flow rate was not addressed.

To determine the effect on gas temperatures, the uncertainty in heptane flow rate was converted to a temperature value using Alpert's correlations for fire plume and ceiling jet temperature rise.[3] To provide a conservative estimate of the uncertainty in temperature resulting from an uncertainty in heptane flow, the uncertainty was calculated by using 20 kW as input to Alpert's correlations. The uncertainty in temperature resulting from an uncertainty of +/- 20 kW would decrease as the heat release rate of the burner increased.

Repeatability uncertainty was estimated by calculating the standard deviation of temperatures measured in replicate tests. Generally, repeatability uncertainty dominated uncertainty from other sources.

The three types of uncertainty in temperature were combined by using the root-sum-of-squares.[9] For purposes of comparing measured temperatures with FDS predictions, experimental data was reported as a range, which was the average of temperatures plus and minus the combined uncertainty. On Figures B.1 – B.96, this data is reported as "high" and low," where "high" is the average temperature plus the combined uncertainty, and "low" is the average temperature minus the combined uncertainty.

An additional source of uncertainty was that the burner may not have been centered directly beneath the thermocouple array that was used to measure temperatures in the plume centerline. If the plume was not centered below the thermocouple array that was used to measure temperatures in the plume centerline, the distances from the plume centerline to the thermocouple arrays that were used to measure temperatures in the ceiling jet region may also have slightly differed from that reported. In the test facility, 100 thermocouples were placed 100 mm below the ceiling in a 3 m by 3 m spacing. The fire source was located below the center of the thermocouple array.

To investigate whether the plume was located at the geometric center of the burner, four thermocouples were selected that were each located at a radial distance of 4.5 meters from the burner (the thermocouples formed a 3 m by 3 m square centered above the burner.)

It was found that the ceiling-level thermocouples did not measure the same temperatures. This demonstrates that the plume was not centered. Additionally, differences in temperature measurements were systematic; they consistently differed throughout a test. Which thermocouples measured higher temperatures varied between tests. There was no variation in which thermocouples measured higher temperatures during a single day, but there was variation from day to day. Because of this, it is suspected that the burner assembly was not exactly centered in the test facility.

However, given that it is not possible to determine exactly where the burner was placed, this source of uncertainty was not addressed. There were not visual obervations recorded in the test documentation relating to the location of the plume.

RESULTS

From the grid convergence studies, it was found that grid size independence was achieved for measurements in the ceiling jet region (radial distances of 2.2, 6.5 and 10.8 meters from the plume centerline.) See Figures B.1-B.16 and B.33-B.48. FDS predictions of temperatures in the plume centerline were found to be sensitive to grid size, and grid size independence was not achieved with the grid spacings that were used in the FDS simulations. Where the plume region ends was not explored in this analysis, since data was only available for detectors located at discrete distances from the plume centerline. However, thermocouples and thermal devices located 2.2 meters from the plume centerline did not exhibit the same sensitivity to grid spacing as did thermocouples and thermal detectors located in the plume centerline.

For the test configuration where the ceiling height was 3.0 meters, FDS predictions of gas and detector temperature were higher than was measured. See Figure 7. Higher predicted detector temperatures would correspond to prediction of detector activation earlier than would be observed.

At a ceiling height of 4.6 meters, FDS predictions were generally within the range of uncertainty outside of the plume region. See Figure 8. At a radial distance of 2.2 meters from the plume centerline, FDS underpredicted the temperature of disks with response time indices of 164 and 287 $m^{1/2}$-$s^{1/2}$. It should be noted that beginning at a time of 218 seconds, the measured plume temperatures in one of the experiments (#02169801) began to decrease, eventually differing by ~100 °C between the two replicate tests. Additionally, measured plume temperatures differed between replicate tests by as much as 75°C during the first 60 seconds. Therefore, the experimental data from these tests should be viewed skeptically.

For the tests with 6.1 and 7.6 meter ceiling heights, FDS predictions were within the range of experimental uncertainty outside of the plume area, although predictions began to fall below the experimental data for devices with response time indexes of 164 and 287 $m^{1/2}$-$s^{1/2}$ and times greater than 300 seconds. See Figure 9 and Figure 10.

At a ceiling height of 10.7 meters, predictions were generally within the range of data, although there were some deviations above and below the range of data. See Figure 11. As the ceiling height increased to 12.2 meters, predictions were within the range of data or above it. See Figure 12

Comparisons of model predictions and experimental data can be found in Appendix B.

ANALYSIS

Overall, predictions outside of the plume region were closer to measured temperatures than within the plume. Part of the reason for this may have been that convergence was not achieved within the plume at the grid spacings (100 mm and 66 mm) used in this study. Given the physical size of the space that was being modeled and the time that it would take to run simulations at finer grid resolutions, further attempts to determine the grid spacing at which convergence would occur were not conducted.

Theoretically, as the grid spacing approaches zero, the results obtained from a numerical solution of a series of partial differential equations will approach the true solution. However, as the grid spacing becomes smaller, the amount of time required for the simulation will increase. The number of grid cells in a simulation will vary with the grid spacing to the third power if the grid spacing in each of the three Cartesian coordinates is uniform.

Therefore, it becomes necessary to balance the desired accuracy with the amount of time available to conduct simulations. In the present study, it was clear that predictions had not converged in the plume region with the grid spacings used. Outside of the plume region, predictions were found to be less sensitive to changes in grid spacing and convergence in the predictions was achieved.

Other work has also investigated the capability of FDS to predict plume temperatures.[10] Predictions of plume temperatures were compared to measurements from experiments that used a 0.9 meter diameter gas fired burner. The referenced investigation also found sensitivity to grid spacing, with the best results occurring at a grid spacing of 50 mm. Additional simulations with smaller grid spacings were not conducted to see if grid convergence had occurred. In lieu of conducting additional simulations with further refined grid spacings, the fact that convergence did not occur in the plume region is noted as a limitation of the present study.

To evaluate the results of the analysis for all scenarios, graphs of predicted temperatures vs. measured temperatures were prepared. In preparing these graphs, predicted and measured temperatures were selected at 100 second intervals. The average of the temperature measurements from replicate experiments at each time interval was plotted, and the calculated uncertainty was displayed using error bars.

Figure 5 contains a plot of data from all ceiling heights, all radial distances, and all types of temperature measurements. The line drawn in Figure 5 shows perfect agreement. Points plotted above the line are cases where FDS predicted higher temperatures than were measured. Points plotted below the line represent cases where FDS predicted lower temperatures than were measured.

As can be seen in Figure 5, many of the temperature predictions are greater than the measured temperatures. However, most of the points where the greatest overpredictions occur are in the plume centerline. Predictions in this area were more sensitive to grid spacing than in other areas, and grid convergence may not have occurred. Other than in Figure 5 and the graphs in Appendix B, measurements and predictions in the plume centerline are not analyzed further. Since grid size convergence was likely not achieved in the plume centerline, the findings in this study should not be applied to predictions of temperature or detector response for locations in the plume region.

Figure 6 shows a plot of data from all ceiling heights and all types of temperature measurements that were taken outside of the plume centerline. This includes all radial distances. The lines drawn in Figure 6 have slopes of 1.9 and 1/1.9. This shows that all predictions outside of the plume centerline were within a factor of 1.9 of the measured temperatures once uncertainty was considered.

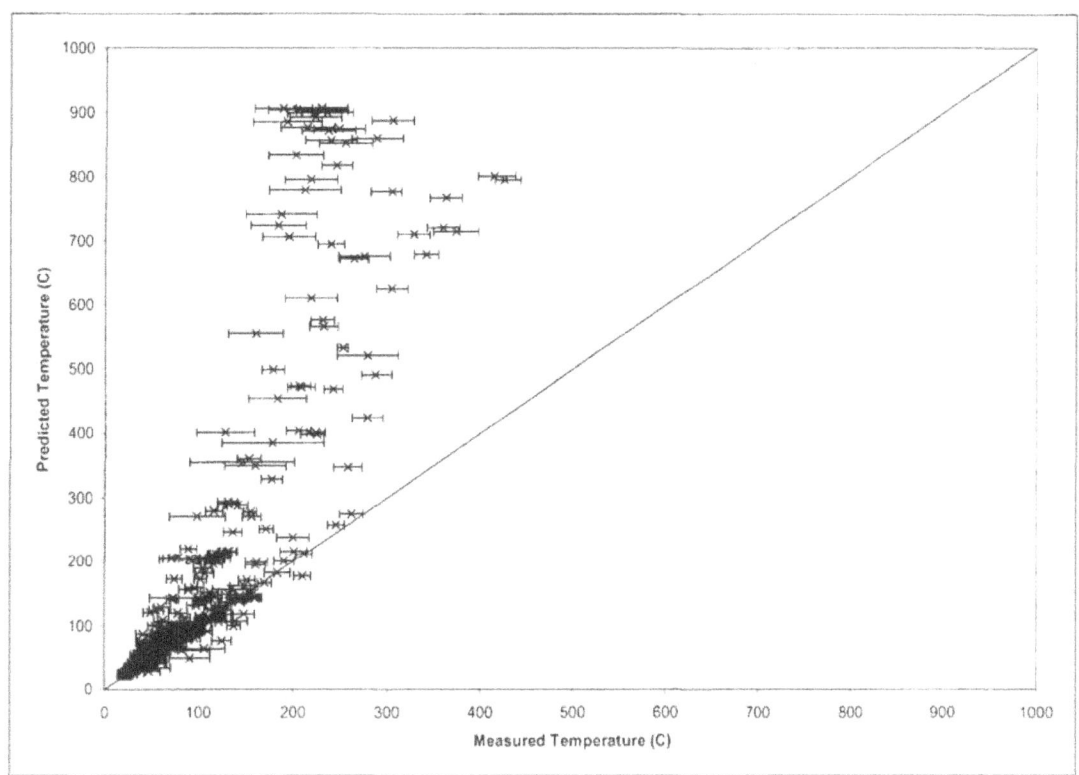

Figure 5 – Comparison of Predicted and Measured Temperatures – All Data

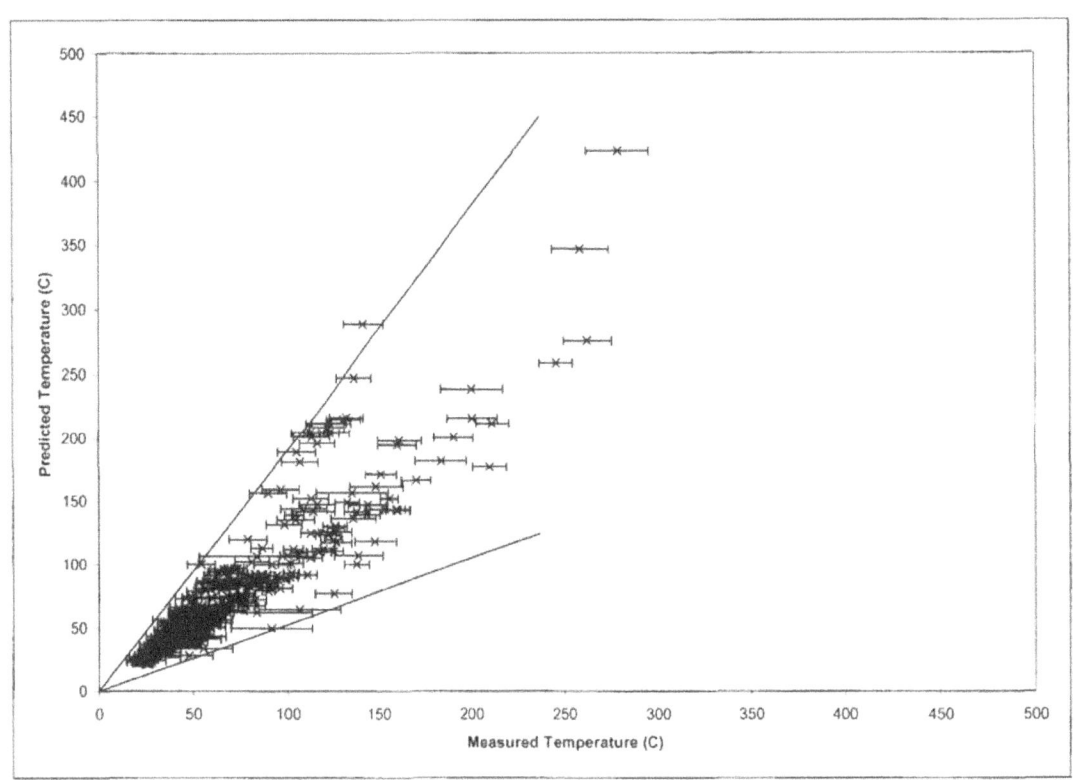

Figure 6 – Comparison of Predicted and Measured Temperatures Outside the Plume Region

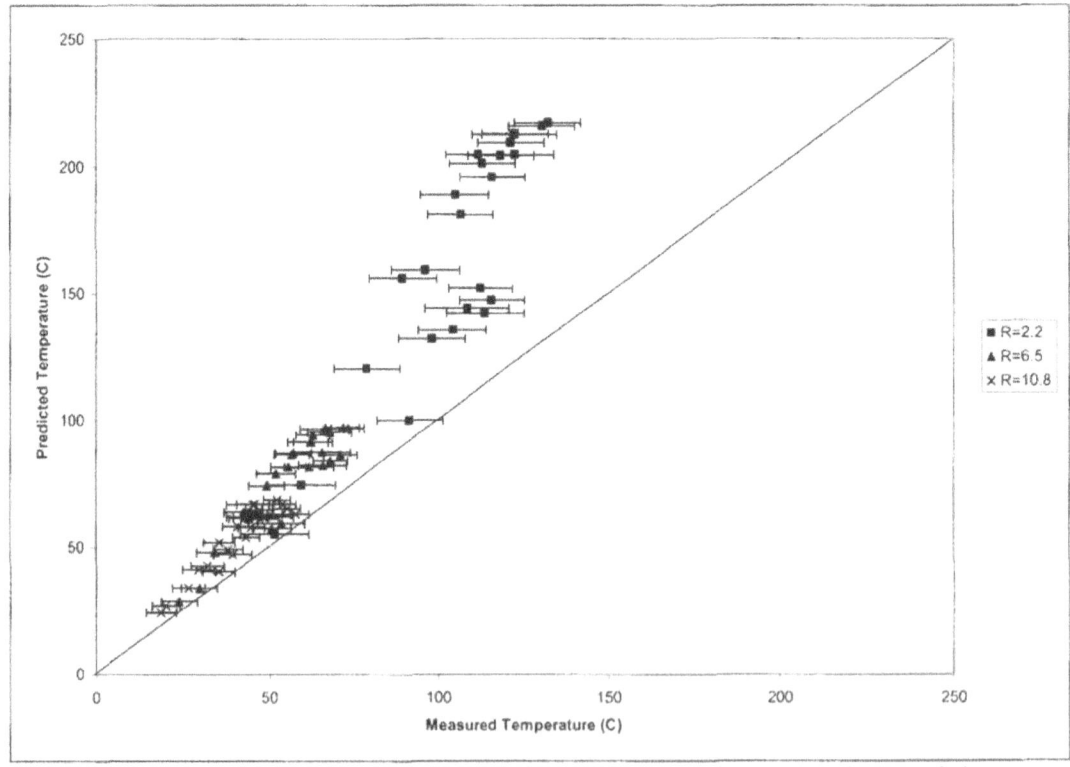

Figure 7 - Comparison of Predicted and Measured Temperatures for H = 3.0 m

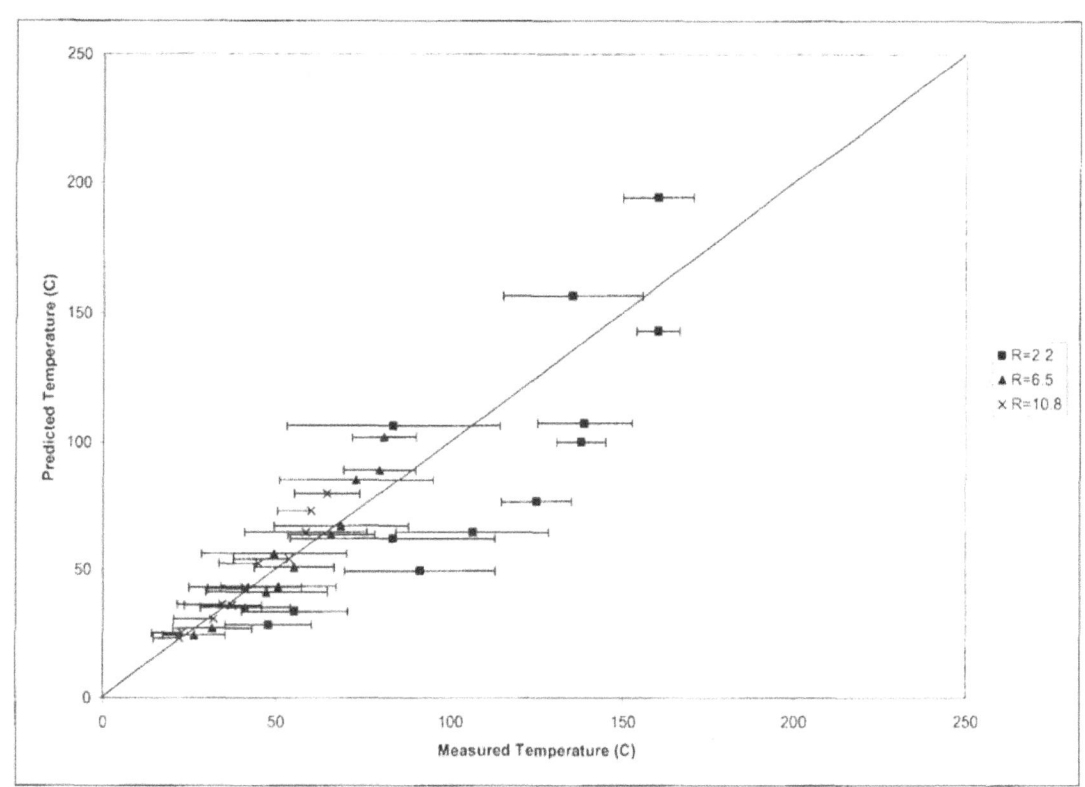

Figure 8 - Comparison of Predicted and Measured Temperatures for H = 4.6 m

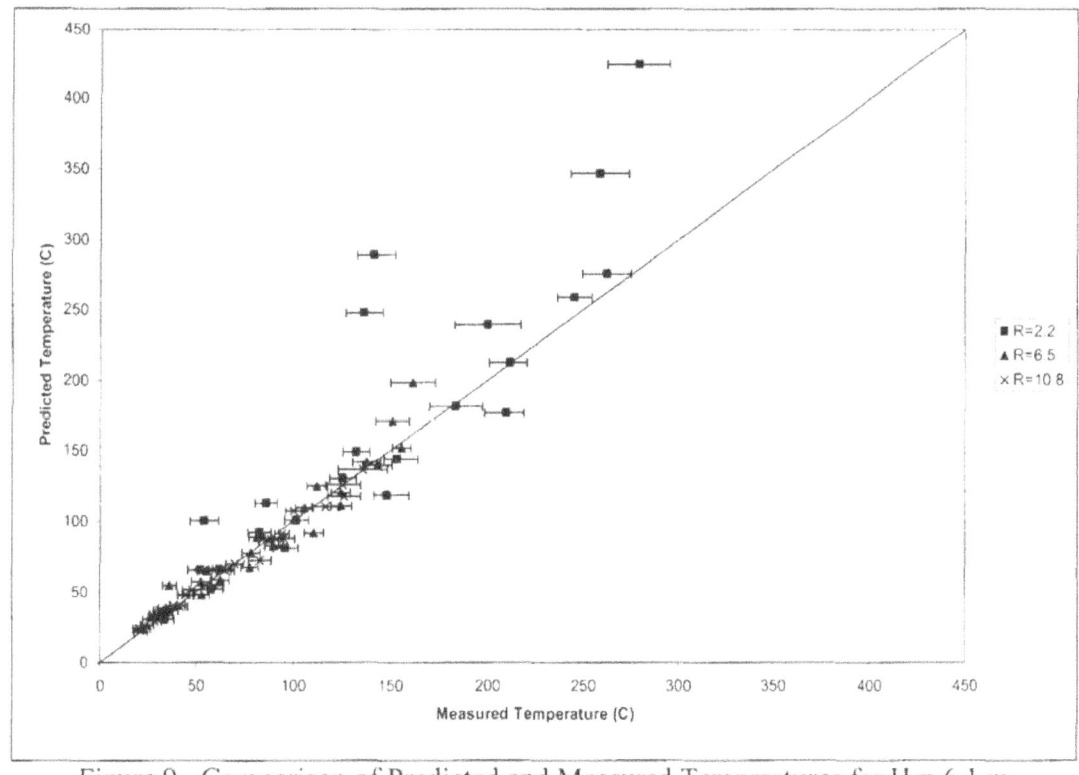

Figure 9 - Comparison of Predicted and Measured Temperatures for H = 6.1 m

13

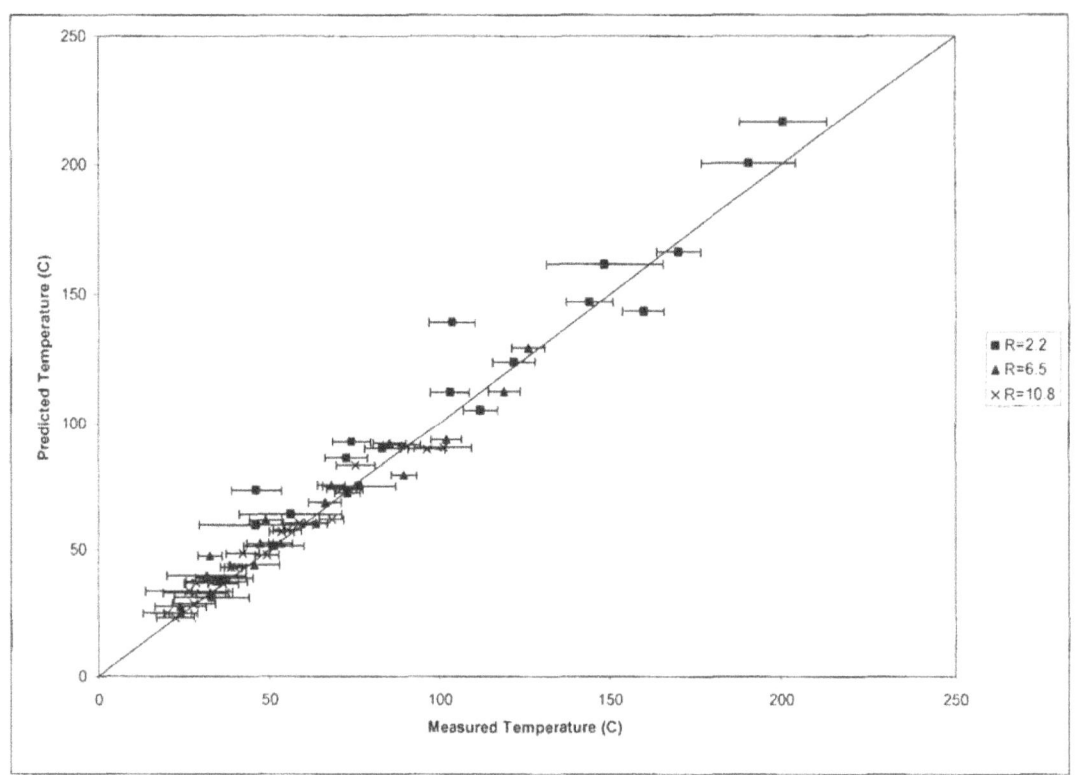

Figure 10 - Comparison of Predicted and Measured Temperatures for H = 7.6 m

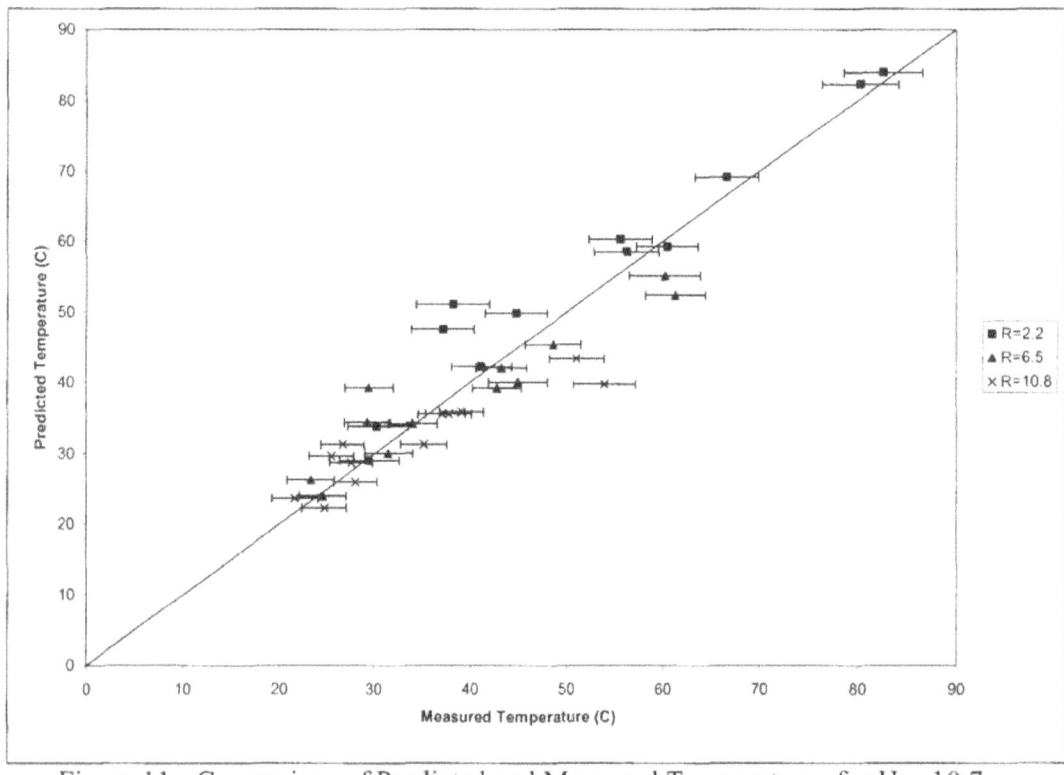

Figure 11 - Comparison of Predicted and Measured Temperatures for H = 10.7 m

14

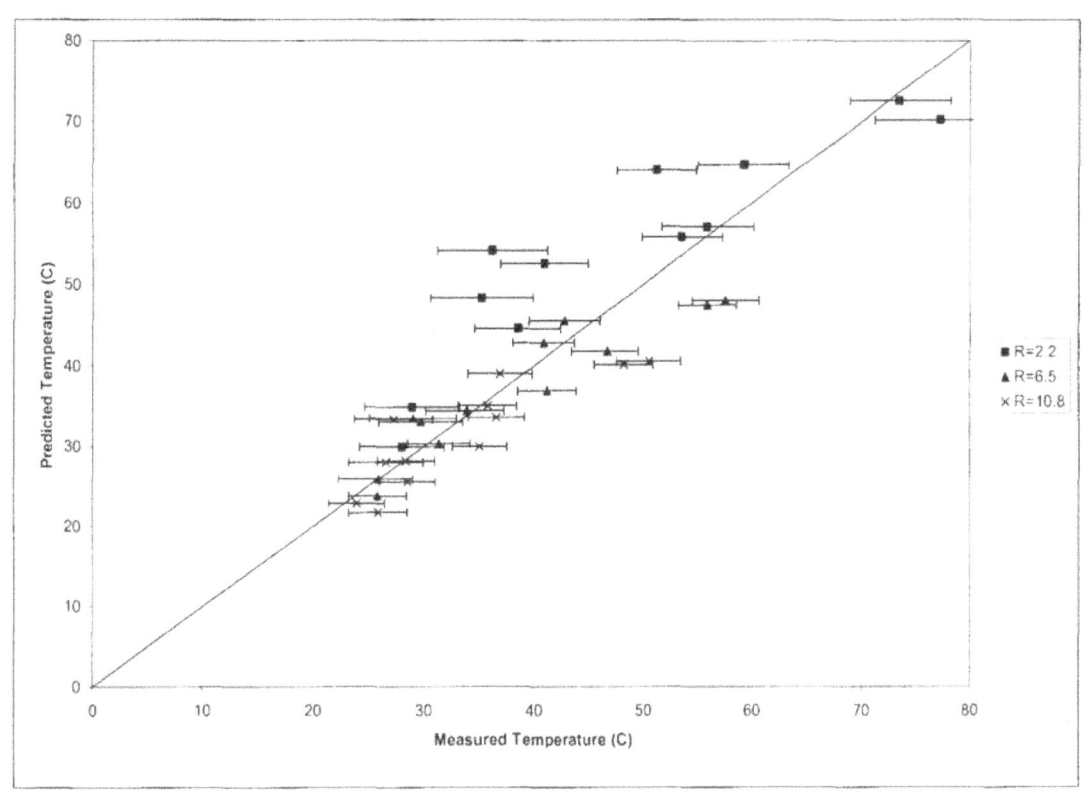

Figure 12 - Comparison of Predicted and Measured Temperatures for H = 12.2 m

It is possible to also look at the influence of ceiling height, radial distance and response time index on the accuracy of predictions. As can be seen in Figure 7 through Figure 12, predictions more closely match measured temperatures as the ceiling height increase.

Figure 13 through Figure 16 show comparisons of predicted and measured thermocouple and heat detector temperatures differentiated by the type of sensing device. As can be seen in these figures, the accuracy of predictions is not strongly influenced by the response time index of the thermal device. However, FDS shows less of a tendency to overpredict temperatures as the response time index increases.

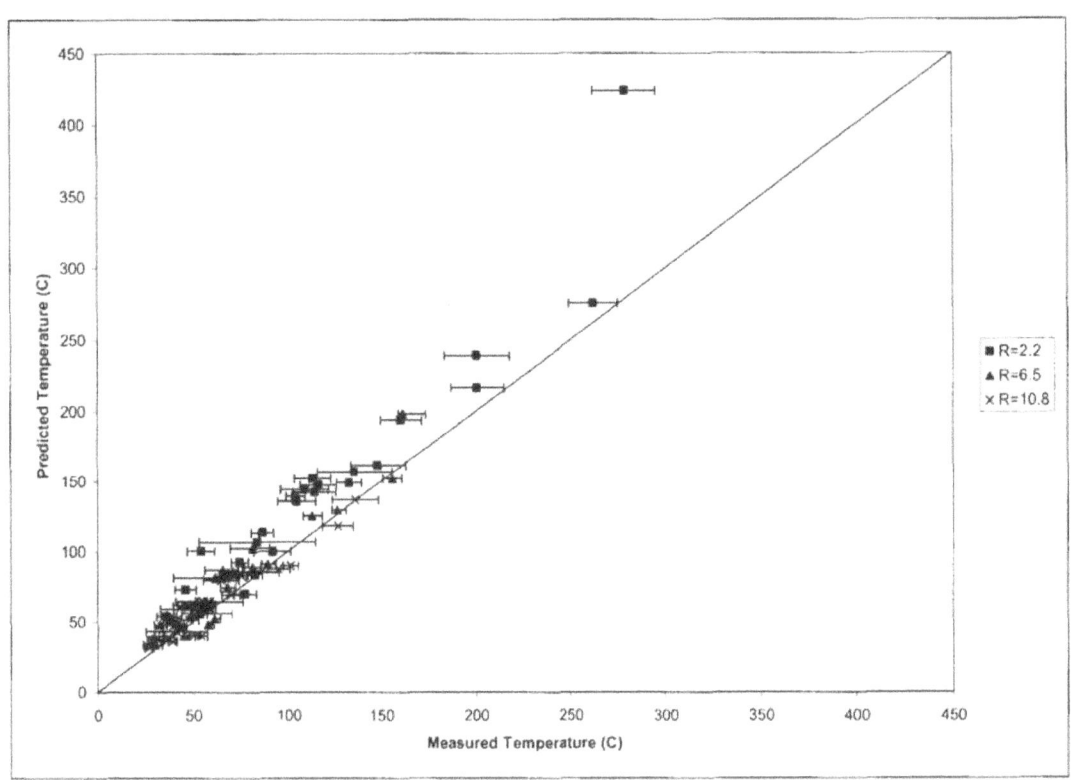

Figure 13 - Comparison of Predicted and Measured Thermocouple (RTI=0) Measurements

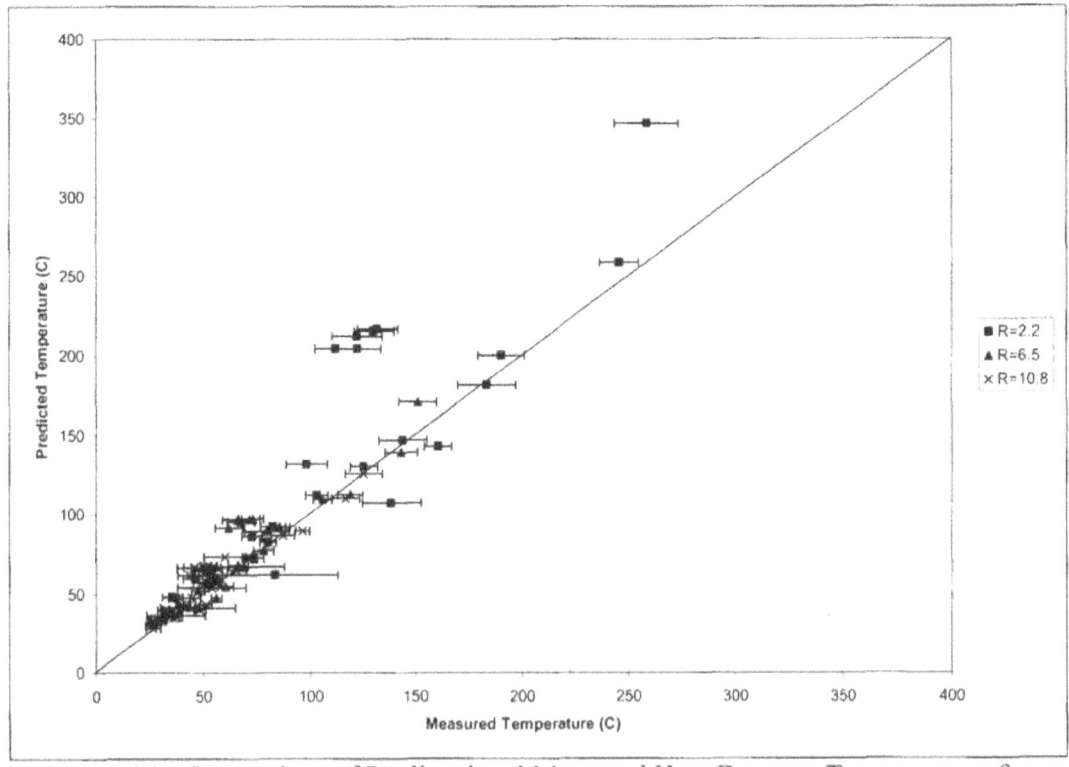

Figure 14 - Comparison of Predicted and Measured Heat Detector Temperatures for
$$RTI = 32 \ m^{1/2}\text{-}s^{1/2}$$

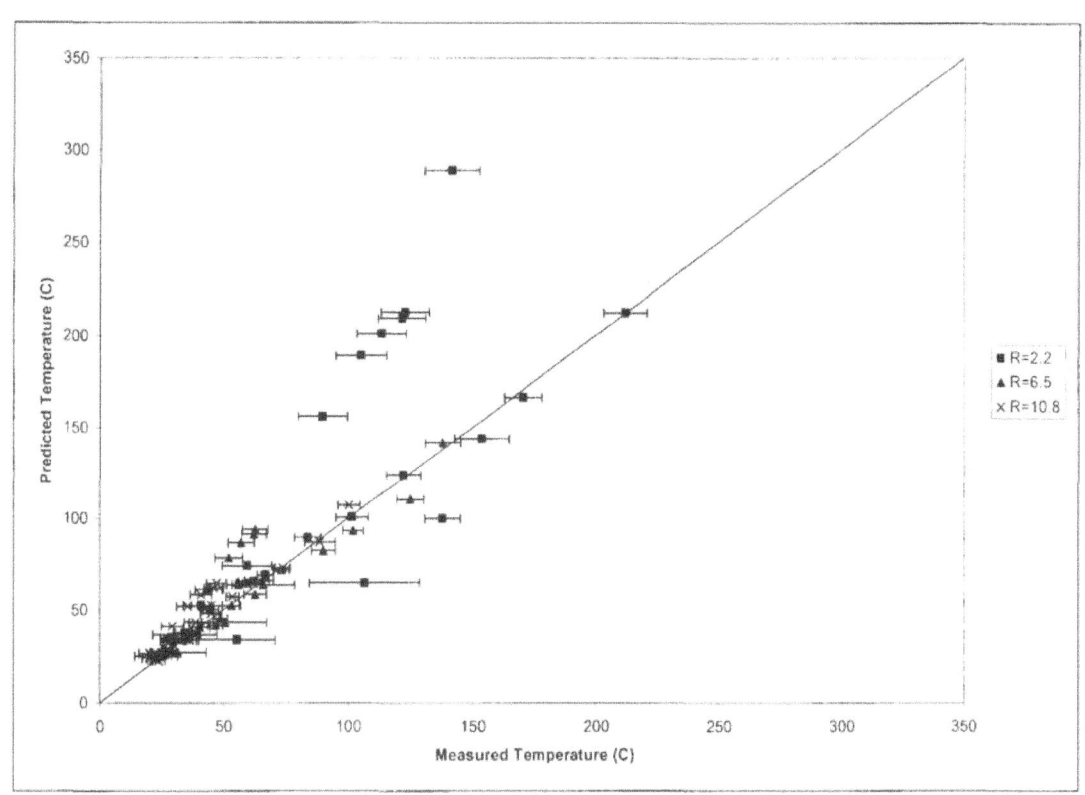

Figure 15 - Comparison of Predicted and Measured Heat Detector Temperatures for RTI = 164 m$^{1/2}$-s$^{1/2}$

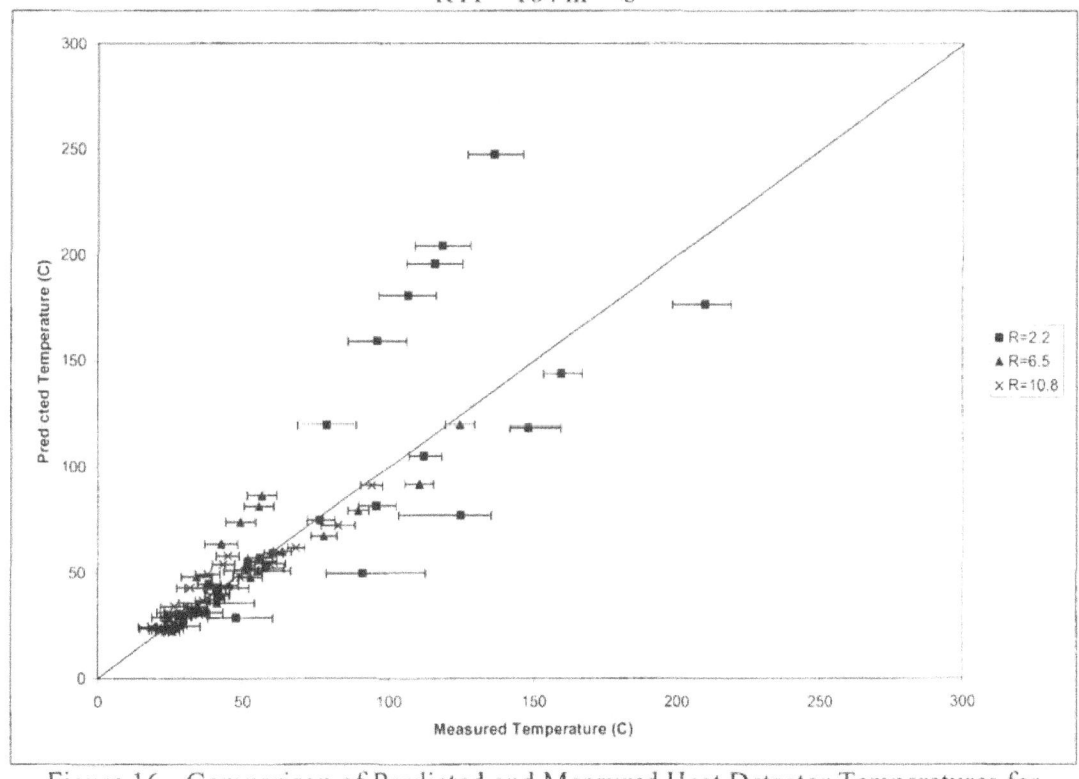

Figure 16 - Comparison of Predicted and Measured Heat Detector Temperatures for RTI = 287 m$^{1/2}$-s$^{1/2}$

17

Figure 17 through Figure 19 shows the influence of radial distance on the accuracy of predictions of temperature. The distance from the plume centerline did not influence the accuracy of FDS predictions.

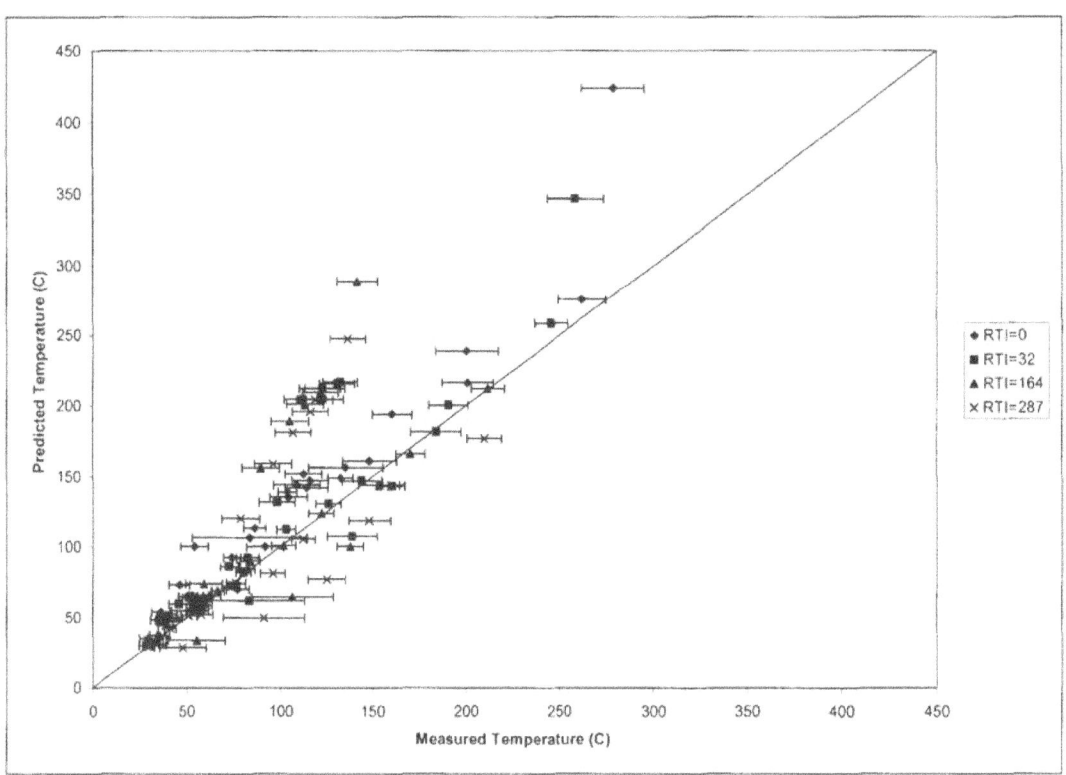

Figure 17 - Comparison of Predicted and Measured Temperatures at R=2.2 m

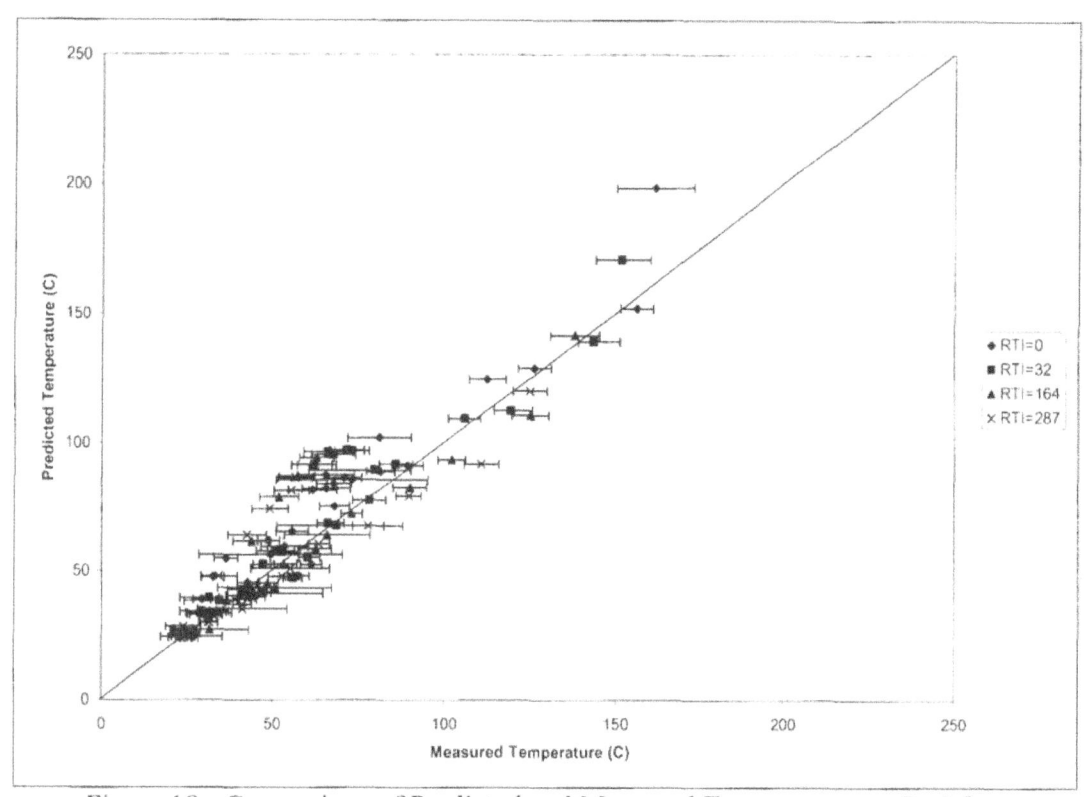

Figure 18 - Comparison of Predicted and Measured Temperatures at R=6.5 m

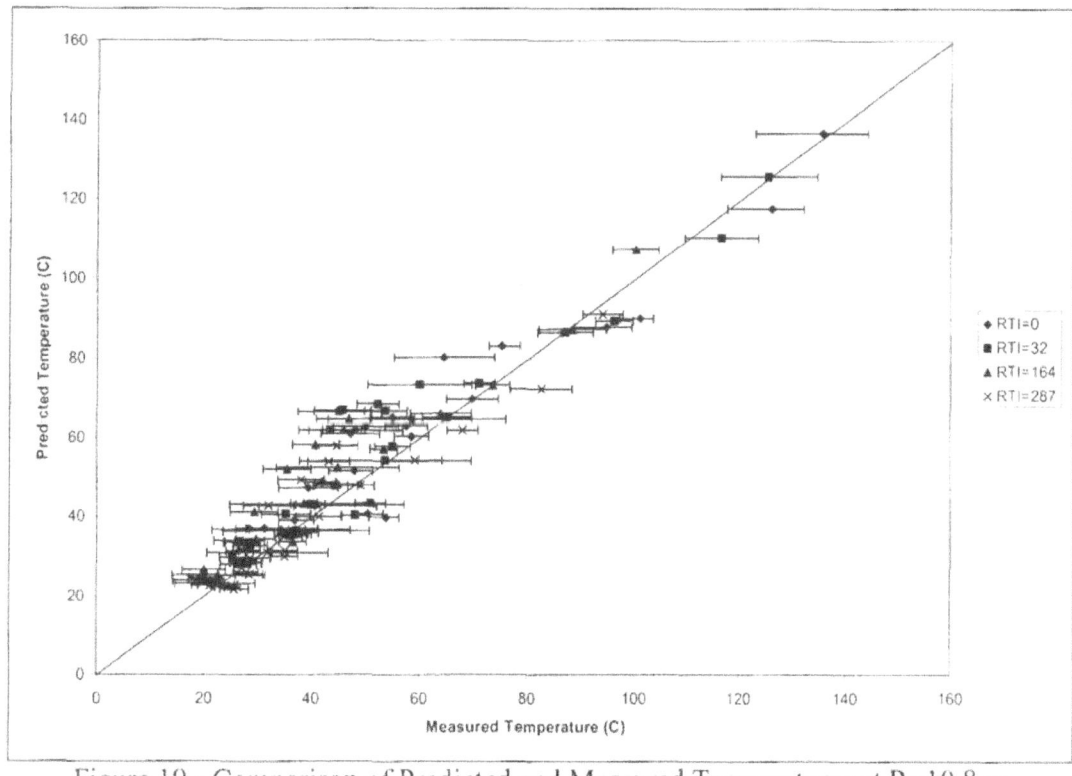

Figure 19 - Comparison of Predicted and Measured Temperatures at R=10.8 m

19

DISCUSSION

Outside of the plume region, FDS provides predictions of gas temperature and thermal detector response within a factor of 1.9 of measurements for fires located under an unobstructed ceiling with ceiling heights ranging from 3.0 to 12.2 meters.

When modeling sprinkler or detector response, the location of the fire is often chosen such that it would yield the slowest possible heat detector or sprinkler response. The fire is typically not placed immediately below the location of a sprinkler or heat detector. Instead, a location that would yield a worst-case activation time is chosen. If the sprinklers or detectors are spaced in a square pattern, the fire would be located a distance of 0.7 multiplied by the sprinkler or detector spacing (measured horizontally). The distance of 2.2 meters from the plume centerline would correspond to a sprinkler or detector spacing of 3.1 meters, which is at the lower end of typical spacings that are used in practice. Therefore, the results from this investigation should be applicable to most cases of interest despite the inability to draw conclusions regarding detectors located in the plume centerline.

It is noteworthy that a verification and validation study performed of FDS by NIST[11] found that predictions of ceiling jet temperature rise were accurate to within a factor of 1.16, and predictions of plume temperatures were accurate to within a factor of 1.14. This analysis did not investigate detector response. However, in the NIST analysis, there were data points that were not bounded by these values.

Also, the NIST study recommended a multiplicative factor of 1.2 for predictions of total heat flux. Again, there were data points that were not bounded by this value. A value of approximately 1.7 would be needed to capture all of the data. See figure 6-11 of reference 11.

All of these multiplicative factors are less than the value of 1.9 is that suggested here. In developing this value, a multiplicative factor was found that bounded at least some portion of the uncertainty range for all of the data used. A smaller multiplicative factor could have been suggested if some outlying data was excluded.

Although the results from this investigation are applicable to the activation of sprinklers, the ability of FDS to model the effect of sprinkler discharge on fire temperatures or fire size was not evaluated. Also, the effect of conduction of heat from sprinklers to sprinkler piping was not considered, since the thermal devices used in the testing would have negligible heat loss to their wiring. FDS does not permit the inclusion of a "C" factor for heat detectors to account for heat loss to the detector mount, and the thermal devices used in the testing were modeled as heat detectors in FDS. A "C" factor can be specified for sprinklers in FDS to estimate heat loss to the sprinkler piping.

CONCLUSION

Outside of the plume region, predicted ceiling jet temperatures and thermal detector temperatures were within a factor of 1.9 of measured temperatures. From a modeling standpoint, "thermal detectors" could represent sprinklers or heat detectors. Therefore, when modeling the activation of sprinklers or heat detectors using FDS, the expected activation time to a given fire should be bounded by predictions using sprinklers or heat detectors that have activation temperatures that are 1.9 times greater and $1/1.9^{th}$ of the activation temperature of the sprinkler or detector being modeled (for temperatures in °C.)

ACKNOWLEDGEMENTS

The authors express their appreciation to Steve Kerber of NIST for assistance provided in FDS modeling.

[1] Engineering Guide – Evaluation of the Computer Fire Model DETACT-QS, Society of Fire Protection Engineers, Bethesda, MD, 2002.

[2] Evans, D. & Stroup, D. "Methods to Calculate the Response Time of Heat and Smoke Detectors Installed Below Large Unobstructed Ceilings," NBSIR 85-3167, National Bureau of Standards, Gaithersburg, MD 1985.

[3] Alpert, R. "Calculation of Response Time of Ceiling-Mounted Fire Detectors, *Fire Technology* **8**:3 (1972) pp. 181-195.

[4] McGrattan, K. & Forney, G. "Fire Dynamics Simulator (Version 4) User's Guide", NIST Special Publication 1019, National Institute of Standards and Technology, Gaithersburg, MD 2005.

[5] "Fire Environment Tests Under Flat Ceilings," Test Report R18476-96NK37932, Underwriters Laboratories, Northbrook, Il., October 1998.

[6] DiNenno, P., Ed. Appendix C, SFPE Handbook of Fire Protection Engineering, National Fire Protection association, Quincy, MA, 2002.

[7] UL 1767, "Standard for Early Suppression Fast Response Sprinklers," Underwriters laboratories, Northbrook, IL, 1995.

[8] Omega Engineering Inc., *The Temperature Handbook*, Vol. MM, pages Z-39-40, Stamford, CT., 2000.

[9] Taylor, B., Kuyatt, C., "Guidelines for Evaluating and Expressing the Uncertainty of NIST Measurement Results," NIST Technical Note 1297, National Institute of Standards and Technology, Gaithersburg, MD, 1994.

[10] Dreisbach, J. & McGrattan, K. "Verification and Validation of Selected Fire Models for Nuclear Power Plant Applications – Volume 6: Fire Dynamics Simulator," NUREG-1824 (Draft for Comment), Nuclear Regulatory Commission & Electric Power Research Institute, Washington, DC, 2006.

[11] McGrattan, K. "Verification and Validation of Selected Fire Models for Nuclear Power Plan Applications – Volume 7: Fire Dynamics Simulator (FDS), NUREG-1824, U.S. Nuclear Regulatory Commission, Washington, DC, 2007.

```
&HEAD CHID='UL10-100mm',TITLE='10 ft ceiling with 100 mm grid
spacing' /

&GRID IBAR=100,JBAR=100,KBAR=30 /
&PDIM XBAR0=10.25, XBAR=20.25, YBAR0=10.25, YBAR=20.25, ZBAR=3.0
/ Plume mesh

&GRID IBAR=100, JBAR=100, KBAR=15/
&PDIM XBAR0=10.25, XBAR=20.25, YBAR0=20.25, YBAR=30.25,
ZBAR0=1.5, ZBAR=3.0/ Ceiling jet mesh

&TIME TWFIN=600 /

&MISC REACTION='HEPTANE', DTCORE=15/

&SURF ID                = 'CEILING TILE'
      RGB               = .95,.95,.95
      FYI               = 'Data as provided by UL'
      KS                = 0.0611
      DENSITY      = 313
      C_P      = 0.753
      DELTA             = 0.0158
      BACKING           = 'EXPOSED'/

&REAC ID='HEPTANE'
      FYI='Heptane, C_7 H_16'
      MW_FUEL=100.
      NU_O2=11.
      NU_CO2=7.
      NU_H2O=8.
      CO_YIELD=0.006
      SOOT_YIELD=0.015 /

*** SPECIFY BOUNDARY CONDITIONS OF MESHES

ALLOW 'FLOOR' OF PLUME MESH TO REMAIN INERT BY DEFAULT

&VENT XB=10.25, 10.25, 10.25, 20.25, 0, 3.00,
SURF_ID='OPEN'/LEFT SIDE OF PLUME MESH
&VENT XB=20.25, 20.25, 10.25, 20.25, 0, 3.00,
SURF_ID='OPEN'/RIGHT SIDE OF PLUME MESH
&VENT XB=10.25, 20.25, 10.25, 10.25, 0, 3.00,
SURF_ID='OPEN'/BOTTOM (YBAR0) OF PLUME MESH
&VENT XB=10.25, 20.25, 10.25, 20.25, 3.00, 3.00,
SURF_ID='CEILING TILE'/CEILING OF PLUME MESH
&VENT XB=10.25, 20.25, 20.25, 20.25, 0, 3.00, SURF_ID='OPEN'/TOP
(YBAR) OF PLUME MESH

&VENT XB=10.25, 20.25, 20.25, 20.25, 1.5, 3.0,
SURF_ID='OPEN'/BOTTOM (YBAR0) OF CEILING JET MESH
&VENT XB=10.25, 10.25, 20.25, 30.25, 1.5, 3.0,
SURF_ID='OPEN'/LEFT SIDE OF CEILING JET MESH
```

```
&VENT XB=20.25, 20.25, 20.25, 30.25, 1.5, 3.0,
SURF_ID='OPEN'/RIGHT SIDE OF CEILING JET MESH
&VENT XB=10.25, 20.25, 30.25, 30.25, 1.5, 3.0,
SURF_ID='OPEN'/TOP (YBAR) OF CEILING JET MESH
&VENT XB=10.25, 20.25, 20.25, 30.25, 1.5, 1.5,
SURF_ID='OPEN'/FLOOR OF CEILING JET MESH
&VENT XB=10.25, 20.25, 20.25, 30.25, 3.0, 3.0, SURF_ID='CEILING
TILE'/CEILING OF CEILING JET MESH

&OBST XB= 14.75,15.75,14.75,15.75,0,0.6, RGB=0,0,1/ burner 1 m x
1 m (39.37 inches by 39.37 inches)

&VENT XB= 14.75,15.75,14.75,15.75,0.6,0.6,VENT_COLOR='RED',
SURF_ID='BURN'/ fire

&SURF ID='BURN',HRRPUA=1055, RAMP_Q='FIRE'/

&RAMP ID='FIRE',  T=0,  F=0.0178/
&RAMP ID='FIRE',  T=5,  F=0.04/
&RAMP ID='FIRE',  T=10,  F=0.0711/
&RAMP ID='FIRE',  T=15,  F=0.1111/
&RAMP ID='FIRE',  T=20,  F=0.16/
&RAMP ID='FIRE',  T=25,  F=0.2177/
&RAMP ID='FIRE',  T=30,  F=0.2844/
&RAMP ID='FIRE',  T=35,  F=0.3599/
&RAMP ID='FIRE',  T=40,  F=0.4436/
&RAMP ID='FIRE',  T=45,  F=0.4661/
&RAMP ID='FIRE',  T=50,  F=0.4891/
&RAMP ID='FIRE',  T=55,  F=0.5126/
&RAMP ID='FIRE',  T=60,  F=0.5368/
&RAMP ID='FIRE',  T=65,  F=0.5614/
&RAMP ID='FIRE',  T=70,  F=0.5867/
&RAMP ID='FIRE',  T=75,  F=0.6124/
&RAMP ID='FIRE',  T=80,  F=0.6388/
&RAMP ID='FIRE',  T=85,  F=0.6657/
&RAMP ID='FIRE',  T=90,  F=0.6931/
&RAMP ID='FIRE',  T=95,  F=0.7211/
&RAMP ID='FIRE',  T=100,  F=0.7497/
&RAMP ID='FIRE',  T=105,  F=0.7788/
&RAMP ID='FIRE',  T=110,  F=0.8085/
&RAMP ID='FIRE',  T=115,  F=0.8387/
&RAMP ID='FIRE',  T=120,  F=0.8695/
&RAMP ID='FIRE',  T=125,  F=0.9008/
&RAMP ID='FIRE',  T=130,  F=0.9327/
&RAMP ID='FIRE',  T=135,  F=0.9651/
&RAMP ID='FIRE',  T=140,  F=0.9981/
&RAMP ID='FIRE',  T=145,  F=1/
&RAMP ID='FIRE',  T=600,  F=1/

&THCP XYZ= 15.25, 15.25, 2.9, QUANTITY='TEMPERATURE',
LABEL='PLUME' /
&HEAT XYZ= 15.25, 15.25, 2.9, RTI=32,
ACTIVATION_TEMPERATURE=1000.0, LABEL='PLUME32'/
&HEAT XYZ= 15.25, 15.25, 2.9, RTI=164,
ACTIVATION_TEMPERATURE=1000, LABEL='PLUME164' /
```

```
&HEAT XYZ= 15.25, 15.25, 2.9, RTI=287,
ACTIVATION_TEMPERATURE=1000.0, LABEL='PLUME287' /

&THCP XYZ= 15.25, 17.41, 2.9, QUANTITY='TEMPERATURE',
LABEL='7FT' /
&HEAT XYZ= 15.25, 17.41, 2.9, RTI=32,
ACTIVATION_TEMPERATURE=1000.0, LABEL='7FT32'/
&HEAT XYZ= 15.25, 17.41, 2.95, RTI=164,
ACTIVATION_TEMPERATURE=1000, LABEL='7FT164' /
&HEAT XYZ= 15.25, 17.41, 2.95, RTI=287,
ACTIVATION_TEMPERATURE=1000.0, LABEL='7FT287' /

&THCP XYZ= 15.25, 21.71, 2.9, QUANTITY='TEMPERATURE',
LABEL='21FT' /
&HEAT XYZ= 15.25, 21.71, 2.9, RTI=32,
ACTIVATION_TEMPERATURE=1000.0, LABEL='21FT32'/
&HEAT XYZ= 15.25, 21.71, 2.9, RTI=164,
ACTIVATION_TEMPERATURE=1000, LABEL='21FT164' /
&HEAT XYZ= 15.25, 21.71, 2.9, RTI=287,
ACTIVATION_TEMPERATURE=1000.0, LABEL='21FT287' /

&THCP XYZ= 15.25, 26.03, 2.9, QUANTITY='TEMPERATURE',
LABEL='35FT' /
&HEAT XYZ= 15.25, 26.03, 2.9, RTI=32,
ACTIVATION_TEMPERATURE=1000.0, LABEL='35FT32'/
&HEAT XYZ= 15.25, 26.03, 2.9, RTI=164,
ACTIVATION_TEMPERATURE=1000, LABEL='35FT164' /
&HEAT XYZ= 15.25, 26.03, 2.9, RTI=287,
ACTIVATION_TEMPERATURE=1000.0, LABEL='35FT287' /

&SLCF PBX=15.25, QUANTITY='TEMPERATURE'/
&SLCF PBX=15.25, QUANTITY='VELOCITY'/
```

```
&HEAD CHID='UL10-66mm',TITLE='10 ft ceiling with 66 mm grid
spacing' /


&GRID IBAR=150,JBAR=150,KBAR=48 /
&PDIM XBAR0=10.25, XBAR=20.25, YBAR0=10.25, YBAR=20.25, ZBAR=3.0
/ Plume mesh

&GRID IBAR=150, JBAR=150, KBAR=24/
&PDIM XBAR0=10.25, XBAR=20.25, YBAR0=20.25, YBAR=30.25,
ZBAR0=1.5, ZBAR=3.0/ Ceiling jet mesh

&TIME TWFIN=600 /

&MISC REACTION='HEPTANE', DTCORE=15/

&SURF ID                      = 'CEILING TILE'
      RGB                     = .95,.95,.95
      FYI                     = 'Data as provided by UL'
      KS                      = 0.0611
      DENSITY           = 313
      C_P          = 0.753
      DELTA                   = 0.0158
      BACKING                 = 'EXPOSED'/

&REAC ID='HEPTANE'
      FYI='Heptane, C_7 H_16'
      MW_FUEL=100.
      NU_O2=11.
      NU_CO2=7.
      NU_H2O=8.
      CO_YIELD=0.006
      SOOT_YIELD=0.015 /

*** SPECIFY BOUNDARY CONDITIONS OF MESHES

ALLOW 'FLOOR' OF PLUME MESH TO REMAIN INERT BY DEFAULT

&VENT XB=10.25, 10.25, 10.25, 20.25, 0, 3.00,
SURF_ID='OPEN'/LEFT SIDE OF PLUME MESH
&VENT XB=20.25, 20.25, 10.25, 20.25, 0, 3.00,
SURF_ID='OPEN'/RIGHT SIDE OF PLUME MESH
&VENT XB=10.25, 20.25, 10.25, 10.25, 0, 3.00,
SURF_ID='OPEN'/BOTTOM (YBAR0) OF PLUME MESH
&VENT XB=10.25, 20.25, 10.25, 20.25, 3.00, 3.00,
SURF_ID='CEILING TILE'/CEILING OF PLUME MESH
&VENT XB=10.25, 20.25, 20.25, 20.25, 0, 3.00, SURF_ID='OPEN'/TOP
(YBAR) OF PLUME MESH

&VENT XB=10.25, 20.25, 20.25, 20.25, 1.5, 3.0,
SURF_ID='OPEN'/BOTTOM (YBAR0) OF CEILING JET MESH
&VENT XB=10.25, 10.25, 20.25, 30.25, 1.5, 3.0,
SURF_ID='OPEN'/LEFT SIDE OF CEILING JET MESH
&VENT XB=20.25, 20.25, 20.25, 30.25, 1.5, 3.0,
SURF_ID='OPEN'/RIGHT SIDE OF CEILING JET MESH
```

```
&VENT XB=10.25, 20.25, 30.25, 30.25, 1.5, 3.0,
SURF_ID='OPEN'/TOP (YBAR) OF CEILING JET MESH
&VENT XB=10.25, 20.25, 20.25, 30.25, 1.5, 1.5,
SURF_ID='OPEN'/FLOOR OF CEILING JET MESH
&VENT XB=10.25, 20.25, 20.25, 30.25, 3.0, 3.0, SURF_ID='CEILING
TILE'/CEILING OF CEILING JET MESH

&OBST XB= 14.75,15.75,14.75,15.75,0,0.6, RGB=0,0,1/ burner 1 m x
1 m (39.37 inches by 39.37 inches)

&VENT XB= 14.75,15.75,14.75,15.75,0.6,0.6,VENT_COLOR='RED',
SURF_ID='BURN'/ fire

&SURF ID='BURN',HRRPUA=1055, RAMP_Q='FIRE'/

&RAMP ID='FIRE', T=0, F=0.0178/
&RAMP ID='FIRE', T=5, F=0.04/
&RAMP ID='FIRE', T=10, F=0.0711/
&RAMP ID='FIRE', T=15, F=0.1111/
&RAMP ID='FIRE', T=20, F=0.16/
&RAMP ID='FIRE', T=25, F=0.2177/
&RAMP ID='FIRE', T=30, F=0.2844/
&RAMP ID='FIRE', T=35, F=0.3599/
&RAMP ID='FIRE', T=40, F=0.4436/
&RAMP ID='FIRE', T=45, F=0.4661/
&RAMP ID='FIRE', T=50, F=0.4891/
&RAMP ID='FIRE', T=55, F=0.5126/
&RAMP ID='FIRE', T=60, F=0.5368/
&RAMP ID='FIRE', T=65, F=0.5614/
&RAMP ID='FIRE', T=70, F=0.5867/
&RAMP ID='FIRE', T=75, F=0.6124/
&RAMP ID='FIRE', T=80, F=0.6388/
&RAMP ID='FIRE', T=85, F=0.6657/
&RAMP ID='FIRE', T=90, F=0.6931/
&RAMP ID='FIRE', T=95, F=0.7211/
&RAMP ID='FIRE', T=100, F=0.7497/
&RAMP ID='FIRE', T=105, F=0.7788/
&RAMP ID='FIRE', T=110, F=0.8085/
&RAMP ID='FIRE', T=115, F=0.8387/
&RAMP ID='FIRE', T=120, F=0.8695/
&RAMP ID='FIRE', T=125, F=0.9008/
&RAMP ID='FIRE', T=130, F=0.9327/
&RAMP ID='FIRE', T=135, F=0.9651/
&RAMP ID='FIRE', T=140, F=0.9981/
&RAMP ID='FIRE', T=145, F=1/
&RAMP ID='FIRE', T=600, F=1/

&THCP XYZ= 15.25, 15.25, 2.9, QUANTITY='TEMPERATURE',
LABEL='PLUME' /
&HEAT XYZ= 15.25, 15.25, 2.9, RTI=32,
ACTIVATION_TEMPERATURE=1000.0, LABEL='PLUME32'/
&HEAT XYZ= 15.25, 15.25, 2.9, RTI=164,
ACTIVATION_TEMPERATURE=1000, LABEL='PLUME164' /
&HEAT XYZ= 15.25, 15.25, 2.9, RTI=287,
ACTIVATION_TEMPERATURE=1000.0, LABEL='PLUME287' /
```

```
&THCP XYZ= 15.25, 17.41, 2.9, QUANTITY='TEMPERATURE',
LABEL='7FT' /
&HEAT XYZ= 15.25, 17.41, 2.9, RTI=32,
ACTIVATION_TEMPERATURE=1000.0, LABEL='7FT32'/
&HEAT XYZ= 15.25, 17.41, 2.95, RTI=164,
ACTIVATION_TEMPERATURE=1000, LABEL='7FT164' /
&HEAT XYZ= 15.25, 17.41, 2.95, RTI=287,
ACTIVATION_TEMPERATURE=1000.0, LABEL='7FT287' /

&THCP XYZ= 15.25, 21.71, 2.9, QUANTITY='TEMPERATURE',
LABEL='21FT' /
&HEAT XYZ= 15.25, 21.71, 2.9, RTI=32,
ACTIVATION_TEMPERATURE=1000.0, LABEL='21FT32'/
&HEAT XYZ= 15.25, 21.71, 2.9, RTI=164,
ACTIVATION_TEMPERATURE=1000, LABEL='21FT164' /
&HEAT XYZ= 15.25, 21.71, 2.9, RTI=287,
ACTIVATION_TEMPERATURE=1000.0, LABEL='21FT287' /

&THCP XYZ= 15.25, 26.03, 2.9, QUANTITY='TEMPERATURE',
LABEL='35FT' /
&HEAT XYZ= 15.25, 26.03, 2.9, RTI=32,
ACTIVATION_TEMPERATURE=1000.0, LABEL='35FT32'/
&HEAT XYZ= 15.25, 26.03, 2.9, RTI=164,
ACTIVATION_TEMPERATURE=1000, LABEL='35FT164' /
&HEAT XYZ= 15.25, 26.03, 2.9, RTI=287,
ACTIVATION_TEMPERATURE=1000.0, LABEL='35FT287' /

&SLCF PBX=15.25, QUANTITY='TEMPERATURE'/
&SLCF PBX=15.25, QUANTITY='VELOCITY'/
```

```
&HEAD CHID='UL15-100mm',TITLE='15 ft ceiling with 100 mm grid
spacing' /


&GRID IBAR=100,JBAR=100,KBAR=48 /
&PDIM XBAR0=10.25, XBAR=20.25, YBAR0=10.25, YBAR=20.25, ZBAR=4.6
/ Plume mesh

&GRID IBAR=100, JBAR=100, KBAR=24/
&PDIM XBAR0=10.25, XBAR=20.25, YBAR0=20.25, YBAR=30.25,
ZBAR0=2.3, ZBAR=4.6/ Ceiling jet mesh

&TIME TWFIN=600 /

&MISC REACTION='HEPTANE', DTCORE=15/

&SURF ID                = 'CEILING TILE'
      RGB               = .95,.95,.95
      FYI               = 'Data as provided by UL'
      KS                = 0.0611
      DENSITY       = 313
      C_P       = 0.753
      DELTA             = 0.0158
      BACKING           = 'EXPOSED'/

&REAC ID='HEPTANE'
      FYI='Heptane, C_7 H_16'
      MW_FUEL=100.
      NU_O2=11.
      NU_CO2=7.
      NU_H2O=8.
      CO_YIELD=0.006
      SOOT_YIELD=0.015 /

*** SPECIFY BOUNDARY CONDITIONS OF MESHES

ALLOW 'FLOOR' OF PLUME MESH TO REMAIN INERT BY DEFAULT

&VENT XB=10.25, 10.25, 10.25, 20.25, 0, 4.6, SURF_ID='OPEN'/LEFT
SIDE OF PLUME MESH
&VENT XB=20.25, 20.25, 10.25, 20.25, 0, 4.6,
SURF_ID='OPEN'/RIGHT SIDE OF PLUME MESH
&VENT XB=10.25, 20.25, 10.25, 10.25, 0, 4.6,
SURF_ID='OPEN'/BOTTOM (YBAR0) OF PLUME MESH
&VENT XB=10.25, 20.25, 10.25, 20.25, 4.6, 4.6, SURF_ID='CEILING
TILE'/CEILING OF PLUME MESH
&VENT XB=10.25, 20.25, 20.25, 20.25, 0, 4.6, SURF_ID='OPEN'/TOP
(YBAR) OF PLUME MESH

&VENT XB=10.25, 20.25, 20.25, 20.25, 2.3, 4.6,
SURF_ID='OPEN'/BOTTOM (YBAR0) OF CEILING JET MESH
&VENT XB=10.25, 10.25, 20.25, 30.25, 2.3, 4.6,
SURF_ID='OPEN'/LEFT SIDE OF CEILING JET MESH
&VENT XB=20.25, 20.25, 20.25, 30.25, 2.3, 4.6,
SURF_ID='OPEN'/RIGHT SIDE OF CEILING JET MESH
```

```
&VENT XB=10.25, 20.25, 30.25, 30.25, 2.3, 4.6,
SURF_ID='OPEN'/TOP (YBAR) OF CEILING JET MESH
&VENT XB=10.25, 20.25, 20.25, 30.25, 2.3, 2.3,
SURF_ID='OPEN'/FLOOR OF CEILING JET MESH
&VENT XB=10.25, 20.25, 20.25, 30.25, 4.6, 4.6, SURF_ID='CEILING
TILE'/CEILING OF CEILING JET MESH

&OBST XB= 14.75,15.75,14.75,15.75,0,0.6, RGB=0,0,1/ burner 1 m x
1 m (39.37 inches by 39.37 inches)

&VENT XB= 14.75,15.75,14.75,15.75,0.6,0.6,VENT_COLOR='RED',
SURF_ID='BURN'/ fire

&SURF ID='BURN',HRRPUA=2100, RAMP_Q='FIRE'/

&RAMP ID='FIRE', T=0, F=0.0089/
&RAMP ID='FIRE', T=5, F=0.0201/
&RAMP ID='FIRE', T=10, F=0.0357/
&RAMP ID='FIRE', T=15, F=0.0558/
&RAMP ID='FIRE', T=20, F=0.0804/
&RAMP ID='FIRE', T=25, F=0.1094/
&RAMP ID='FIRE', T=30, F=0.1429/
&RAMP ID='FIRE', T=35, F=0.1808/
&RAMP ID='FIRE', T=40, F=0.2229/
&RAMP ID='FIRE', T=45, F=0.2341/
&RAMP ID='FIRE', T=50, F=0.2457/
&RAMP ID='FIRE', T=55, F=0.2575/
&RAMP ID='FIRE', T=60, F=0.2697/
&RAMP ID='FIRE', T=65, F=0.2821/
&RAMP ID='FIRE', T=70, F=0.2947/
&RAMP ID='FIRE', T=75, F=0.3077/
&RAMP ID='FIRE', T=80, F=0.3209/
&RAMP ID='FIRE', T=85, F=0.3344/
&RAMP ID='FIRE', T=90, F=0.3482/
&RAMP ID='FIRE', T=95, F=0.3623/
&RAMP ID='FIRE', T=100, F=0.3766/
&RAMP ID='FIRE', T=105, F=0.3913/
&RAMP ID='FIRE', T=110, F=0.4062/
&RAMP ID='FIRE', T=115, F=0.4213/
&RAMP ID='FIRE', T=120, F=0.4368/
&RAMP ID='FIRE', T=125, F=0.4525/
&RAMP ID='FIRE', T=130, F=0.4686/
&RAMP ID='FIRE', T=135, F=0.4849/
&RAMP ID='FIRE', T=140, F=0.5014/
&RAMP ID='FIRE', T=145, F=0.5183/
&RAMP ID='FIRE', T=150, F=0.5354/
&RAMP ID='FIRE', T=155, F=0.5528/
&RAMP ID='FIRE', T=160, F=0.5705/
&RAMP ID='FIRE', T=165, F=0.5885/
&RAMP ID='FIRE', T=170, F=0.6067/
&RAMP ID='FIRE', T=175, F=0.6253/
&RAMP ID='FIRE', T=180, F=0.6441/
&RAMP ID='FIRE', T=185, F=0.6631/
&RAMP ID='FIRE', T=190, F=0.6825/
&RAMP ID='FIRE', T=195, F=0.7021/
```

```
&RAMP ID='FIRE', T=200, F=0.7221/
&RAMP ID='FIRE', T=205, F=0.7423/
&RAMP ID='FIRE', T=210, F=0.7627/
&RAMP ID='FIRE', T=215, F=0.7835/
&RAMP ID='FIRE', T=220, F=0.8045/
&RAMP ID='FIRE', T=225, F=0.8258/
&RAMP ID='FIRE', T=230, F=0.8474/
&RAMP ID='FIRE', T=235, F=0.8693/
&RAMP ID='FIRE', T=240, F=0.8914/
&RAMP ID='FIRE', T=245, F=0.9139/
&RAMP ID='FIRE', T=250, F=0.9366/
&RAMP ID='FIRE', T=255, F=0.9595/
&RAMP ID='FIRE', T=260, F=0.9828/
&RAMP ID='FIRE', T=265, F=1/
&RAMP ID='FIRE', T=600, F=1/


&THCP XYZ= 15.25, 15.25, 4.5, QUANTITY='TEMPERATURE',
LABEL='PLUME' /
&HEAT XYZ= 15.25, 15.25, 4.5, RTI=32,
ACTIVATION_TEMPERATURE=1000.0, LABEL='PLUME32'/
&HEAT XYZ= 15.25, 15.25, 4.5, RTI=164,
ACTIVATION_TEMPERATURE=1000, LABEL='PLUME164' /
&HEAT XYZ= 15.25, 15.25, 4.5, RTI=287,
ACTIVATION_TEMPERATURE=1000.0, LABEL='PLUME287' /

&THCP XYZ= 15.25, 17.41, 4.5, QUANTITY='TEMPERATURE',
LABEL='7FT' /
&HEAT XYZ= 15.25, 17.41, 4.5, RTI=32,
ACTIVATION_TEMPERATURE=1000.0, LABEL='7FT32'/
&HEAT XYZ= 15.25, 17.41, 4.5, RTI=164,
ACTIVATION_TEMPERATURE=1000, LABEL='7FT164' /
&HEAT XYZ= 15.25, 17.41, 4.5, RTI=287,
ACTIVATION_TEMPERATURE=1000.0, LABEL='7FT287' /

&THCP XYZ= 15.25, 21.71, 4.5, QUANTITY='TEMPERATURE',
LABEL='21FT' /
&HEAT XYZ= 15.25, 21.71, 4.5, RTI=32,
ACTIVATION_TEMPERATURE=1000.0, LABEL='21FT32'/
&HEAT XYZ= 15.25, 21.71, 4.5, RTI=164,
ACTIVATION_TEMPERATURE=1000, LABEL='21FT164' /
&HEAT XYZ= 15.25, 21.71, 4.5, RTI=287,
ACTIVATION_TEMPERATURE=1000.0, LABEL='21FT287' /

&THCP XYZ= 15.25, 26.03, 4.5, QUANTITY='TEMPERATURE',
LABEL='35FT' /
&HEAT XYZ= 15.25, 26.03, 4.5, RTI=32,
ACTIVATION_TEMPERATURE=1000.0, LABEL='35FT32'/
&HEAT XYZ= 15.25, 26.03, 4.5, RTI=164,
ACTIVATION_TEMPERATURE=1000, LABEL='35FT164' /
&HEAT XYZ= 15.25, 26.03, 4.5, RTI=287,
ACTIVATION_TEMPERATURE=1000.0, LABEL='35FT287' /

&SLCF PBX=15.25, QUANTITY='TEMPERATURE'/
&SLCF PBX=15.25, QUANTITY='VELOCITY'/
```

```
&HEAD CHID='UL20-100mm',TITLE='20 ft ceiling with 100 mm grid
spacing' /


&GRID IBAR=100,JBAR=100,KBAR=64 /
&PDIM XBAR0=10.25, XBAR=20.25, YBAR0=10.25, YBAR=20.25, ZBAR=6.1
/ Plume mesh

&GRID IBAR=100, JBAR=100, KBAR=32/
&PDIM XBAR0=10.25, XBAR=20.25, YBAR0=20.25, YBAR=30.25,
ZBAR0=3.05, ZBAR=6.1/ Ceiling jet mesh

&TIME TWFIN=600 /

&MISC REACTION='HEPTANE', DTCORE=15/

&SURF ID                  = 'CEILING TILE'
      RGB                 = .95,.95,.95
      FYI                 = 'Data as provided by UL'
      KS                  = 0.0611
      DENSITY       = 313
      C_P       = 0.753
      DELTA               = 0.0158
      BACKING             = 'EXPOSED'/

&REAC ID='HEPTANE'
      FYI='Heptane, C_7 H_16'
      MW_FUEL=100.
      NU_O2=11.
      NU_CO2=7.
      NU_H2O=8.
      CO_YIELD=0.006
      SOOT_YIELD=0.015 /

*** SPECIFY BOUNDARY CONDITIONS OF MESHES

ALLOW 'FLOOR' OF PLUME MESH TO REMAIN INERT BY DEFAULT

&VENT XB=10.25, 10.25, 10.25, 20.25, 0, 6.1, SURF_ID='OPEN'/LEFT
SIDE OF PLUME MESH
&VENT XB=20.25, 20.25, 10.25, 20.25, 0, 6.1,
SURF_ID='OPEN'/RIGHT SIDE OF PLUME MESH
&VENT XB=10.25, 20.25, 10.25, 10.25, 0, 6.1,
SURF_ID='OPEN'/BOTTOM (YBAR0) OF PLUME MESH
&VENT XB=10.25, 20.25, 10.25, 20.25, 6.1, 6.1, SURF_ID='CEILING
TILE'/CEILING OF PLUME MESH
&VENT XB=10.25, 20.25, 20.25, 20.25, 0, 6.1, SURF_ID='OPEN'/TOP
(YBAR) OF PLUME MESH

&VENT XB=10.25, 20.25, 20.25, 20.25, 3.05, 6.1,
SURF_ID='OPEN'/BOTTOM (YBAR0) OF CEILING JET MESH
&VENT XB=10.25, 10.25, 20.25, 30.25, 3.05, 6.1,
SURF_ID='OPEN'/LEFT SIDE OF CEILING JET MESH
&VENT XB=20.25, 20.25, 20.25, 30.25, 3.05, 6.1,
SURF_ID='OPEN'/RIGHT SIDE OF CEILING JET MESH
```

&VENT XB=10.25, 20.25, 30.25, 30.25, 3.05, 6.1,
SURF_ID='OPEN'/TOP (YBAR) OF CEILING JET MESH
&VENT XB=10.25, 20.25, 20.25, 30.25, 3.05, 3.05,
SURF_ID='OPEN'/FLOOR OF CEILING JET MESH
&VENT XB=10.25, 20.25, 20.25, 30.25, 6.1, 6.1, SURF_ID='CEILING
TILE'/CEILING OF CEILING JET MESH

&OBST XB= 14.75,15.75,14.75,15.75,0,0.6, RGB=0,0,1/ burner 1 m x
1 m (39.37 inches by 39.37 inches)

&VENT XB= 14.75,15.75,14.75,15.75,0.6,0.6,VENT_COLOR='RED',
SURF_ID='BURN'/ fire

&SURF ID='BURN',HRRPUA=7500, RAMP_Q='FIRE'/

&RAMP ID='FIRE', T=0, F=0.0025/
&RAMP ID='FIRE', T=5, F=0.0056/
&RAMP ID='FIRE', T=10, F=0.01/
&RAMP ID='FIRE', T=15, F=0.0156/
&RAMP ID='FIRE', T=20, F=0.0225/
&RAMP ID='FIRE', T=25, F=0.0306/
&RAMP ID='FIRE', T=30, F=0.04/
&RAMP ID='FIRE', T=35, F=0.0506/
&RAMP ID='FIRE', T=40, F=0.0624/
&RAMP ID='FIRE', T=45, F=0.0656/
&RAMP ID='FIRE', T=50, F=0.0688/
&RAMP ID='FIRE', T=55, F=0.0721/
&RAMP ID='FIRE', T=60, F=0.0755/
&RAMP ID='FIRE', T=65, F=0.079/
&RAMP ID='FIRE', T=70, F=0.0825/
&RAMP ID='FIRE', T=75, F=0.0862/
&RAMP ID='FIRE', T=80, F=0.0899/
&RAMP ID='FIRE', T=85, F=0.0936/
&RAMP ID='FIRE', T=90, F=0.0975/
&RAMP ID='FIRE', T=95, F=0.1014/
&RAMP ID='FIRE', T=100, F=0.1055/
&RAMP ID='FIRE', T=105, F=0.1096/
&RAMP ID='FIRE', T=110, F=0.1137/
&RAMP ID='FIRE', T=115, F=0.118/
&RAMP ID='FIRE', T=120, F=0.1223/
&RAMP ID='FIRE', T=125, F=0.1267/
&RAMP ID='FIRE', T=130, F=0.1312/
&RAMP ID='FIRE', T=135, F=0.1358/
&RAMP ID='FIRE', T=140, F=0.1404/
&RAMP ID='FIRE', T=145, F=0.1451/
&RAMP ID='FIRE', T=150, F=0.1499/
&RAMP ID='FIRE', T=155, F=0.1548/
&RAMP ID='FIRE', T=160, F=0.1597/
&RAMP ID='FIRE', T=165, F=0.1648/
&RAMP ID='FIRE', T=170, F=0.1699/
&RAMP ID='FIRE', T=175, F=0.1751/
&RAMP ID='FIRE', T=180, F=0.1803/
&RAMP ID='FIRE', T=185, F=0.1857/
&RAMP ID='FIRE', T=190, F=0.1911/
&RAMP ID='FIRE', T=195, F=0.1966/

```
&RAMP ID='FIRE',  T=200,  F=0.2022/
&RAMP ID='FIRE',  T=205,  F=0.2078/
&RAMP ID='FIRE',  T=210,  F=0.2136/
&RAMP ID='FIRE',  T=215,  F=0.2194/
&RAMP ID='FIRE',  T=220,  F=0.2253/
&RAMP ID='FIRE',  T=225,  F=0.2312/
&RAMP ID='FIRE',  T=230,  F=0.2373/
&RAMP ID='FIRE',  T=235,  F=0.2434/
&RAMP ID='FIRE',  T=240,  F=0.2496/
&RAMP ID='FIRE',  T=245,  F=0.2559/
&RAMP ID='FIRE',  T=250,  F=0.2622/
&RAMP ID='FIRE',  T=255,  F=0.2687/
&RAMP ID='FIRE',  T=260,  F=0.2752/
&RAMP ID='FIRE',  T=265,  F=0.2818/
&RAMP ID='FIRE',  T=270,  F=0.2884/
&RAMP ID='FIRE',  T=275,  F=0.2952/
&RAMP ID='FIRE',  T=280,  F=0.302/
&RAMP ID='FIRE',  T=285,  F=0.3089/
&RAMP ID='FIRE',  T=290,  F=0.3159/
&RAMP ID='FIRE',  T=295,  F=0.323/
&RAMP ID='FIRE',  T=300,  F=0.3301/
&RAMP ID='FIRE',  T=305,  F=0.3373/
&RAMP ID='FIRE',  T=310,  F=0.3446/
&RAMP ID='FIRE',  T=315,  F=0.352/
&RAMP ID='FIRE',  T=320,  F=0.3594/
&RAMP ID='FIRE',  T=325,  F=0.367/
&RAMP ID='FIRE',  T=330,  F=0.3746/
&RAMP ID='FIRE',  T=335,  F=0.3822/
&RAMP ID='FIRE',  T=340,  F=0.39/
&RAMP ID='FIRE',  T=345,  F=0.3978/
&RAMP ID='FIRE',  T=350,  F=0.4058/
&RAMP ID='FIRE',  T=355,  F=0.4138/
&RAMP ID='FIRE',  T=360,  F=0.4218/
&RAMP ID='FIRE',  T=365,  F=0.43/
&RAMP ID='FIRE',  T=370,  F=0.4382/
&RAMP ID='FIRE',  T=375,  F=0.4465/
&RAMP ID='FIRE',  T=380,  F=0.4549/
&RAMP ID='FIRE',  T=385,  F=0.4634/
&RAMP ID='FIRE',  T=390,  F=0.4719/
&RAMP ID='FIRE',  T=395,  F=0.4805/
&RAMP ID='FIRE',  T=400,  F=0.4892/
&RAMP ID='FIRE',  T=405,  F=0.498/
&RAMP ID='FIRE',  T=410,  F=0.5068/
&RAMP ID='FIRE',  T=415,  F=0.5158/
&RAMP ID='FIRE',  T=420,  F=0.5248/
&RAMP ID='FIRE',  T=425,  F=0.5339/
&RAMP ID='FIRE',  T=430,  F=0.543/
&RAMP ID='FIRE',  T=435,  F=0.5523/
&RAMP ID='FIRE',  T=440,  F=0.5616/
&RAMP ID='FIRE',  T=445,  F=0.571/
&RAMP ID='FIRE',  T=450,  F=0.5805/
&RAMP ID='FIRE',  T=455,  F=0.59/
&RAMP ID='FIRE',  T=460,  F=0.5997/
&RAMP ID='FIRE',  T=465,  F=0.6094/
&RAMP ID='FIRE',  T=470,  F=0.6192/
```

```
&RAMP ID='FIRE', T=475, F=0.629/
&RAMP ID='FIRE', T=480, F=0.639/
&RAMP ID='FIRE', T=485, F=0.649/
&RAMP ID='FIRE', T=490, F=0.6591/
&RAMP ID='FIRE', T=495, F=0.6693/
&RAMP ID='FIRE', T=500, F=0.6795/
&RAMP ID='FIRE', T=505, F=0.6899/
&RAMP ID='FIRE', T=510, F=0.7003/
&RAMP ID='FIRE', T=515, F=0.7108/
&RAMP ID='FIRE', T=520, F=0.7213/
&RAMP ID='FIRE', T=525, F=0.732/
&RAMP ID='FIRE', T=530, F=0.7427/
&RAMP ID='FIRE', T=535, F=0.7535/
&RAMP ID='FIRE', T=540, F=0.7644/
&RAMP ID='FIRE', T=545, F=0.7754/
&RAMP ID='FIRE', T=550, F=0.7864/
&RAMP ID='FIRE', T=555, F=0.7975/
&RAMP ID='FIRE', T=560, F=0.8087/
&RAMP ID='FIRE', T=565, F=0.82/
&RAMP ID='FIRE', T=570, F=0.8313/
&RAMP ID='FIRE', T=575, F=0.8428/
&RAMP ID='FIRE', T=580, F=0.8543/
&RAMP ID='FIRE', T=585, F=0.8658/
&RAMP ID='FIRE', T=590, F=0.8775/
&RAMP ID='FIRE', T=595, F=0.8892/
&RAMP ID='FIRE', T=600, F=0.9011/

&THCP XYZ= 15.25, 15.25, 6, QUANTITY='TEMPERATURE',
LABEL='PLUME' /
&HEAT XYZ= 15.25, 15.25, 6, RTI=32,
ACTIVATION_TEMPERATURE=1000.0, LABEL='PLUME32'/
&HEAT XYZ= 15.25, 15.25, 6, RTI=164,
ACTIVATION_TEMPERATURE=1000, LABEL='PLUME164' /
&HEAT XYZ= 15.25, 15.25, 6, RTI=287,
ACTIVATION_TEMPERATURE=1000.0, LABEL='PLUME287' /

&THCP XYZ= 15.25, 17.41, 6, QUANTITY='TEMPERATURE', LABEL='7FT'
/
&HEAT XYZ= 15.25, 17.41, 6, RTI=32,
ACTIVATION_TEMPERATURE=1000.0, LABEL='7FT32'/
&HEAT XYZ= 15.25, 17.41, 6, RTI=164,
ACTIVATION_TEMPERATURE=1000, LABEL='7FT164' /
&HEAT XYZ= 15.25, 17.41, 6, RTI=287,
ACTIVATION_TEMPERATURE=1000.0, LABEL='7FT287' /

&THCP XYZ= 15.25, 21.71, 6, QUANTITY='TEMPERATURE', LABEL='21FT'
/
&HEAT XYZ= 15.25, 21.71, 6, RTI=32,
ACTIVATION_TEMPERATURE=1000.0, LABEL='21FT32'/
&HEAT XYZ= 15.25, 21.71, 6, RTI=164,
ACTIVATION_TEMPERATURE=1000, LABEL='21FT164' /
&HEAT XYZ= 15.25, 21.71, 6, RTI=287,
ACTIVATION_TEMPERATURE=1000.0, LABEL='21FT287' /
```

```
&THCP XYZ= 15.25, 26.03, 6, QUANTITY='TEMPERATURE', LABEL='35FT'
/
&HEAT XYZ= 15.25, 26.03, 6, RTI=32,
ACTIVATION_TEMPERATURE=1000.0, LABEL='35FT32'/
&HEAT XYZ= 15.25, 26.03, 6, RTI=164,
ACTIVATION_TEMPERATURE=1000, LABEL='35FT164' /
&HEAT XYZ= 15.25, 26.03, 6, RTI=287,
ACTIVATION_TEMPERATURE=1000.0, LABEL='35FT287' /

&SLCF PBX=15.25, QUANTITY='TEMPERATURE'/
&SLCF PBX=15.25, QUANTITY='VELOCITY'/
```

```
&HEAD CHID='UL20-66mm',TITLE='20 ft ceiling with 66 mm grid
spacing' /


&GRID IBAR=150,JBAR=150,KBAR=90 /
&PDIM XBAR0=10.25, XBAR=20.25, YBAR0=10.25, YBAR=20.25, ZBAR=6.1
/ Plume mesh

&GRID IBAR=150, JBAR=150, KBAR=45/
&PDIM XBAR0=10.25, XBAR=20.25, YBAR0=20.25, YBAR=30.25,
ZBAR0=3.05, ZBAR=6.1/ Ceiling jet mesh

&TIME TWFIN=600 /

&MISC REACTION='HEPTANE', DTCORE=15/

&SURF ID                  = 'CEILING TILE'
      RGB                 = .95,.95,.95
      FYI                 = 'Data as provided by UL'
      KS                  = 0.0611
      DENSITY       = 313
      C_P        = 0.753
      DELTA               = 0.0158
      BACKING             = 'EXPOSED'/

&REAC ID='HEPTANE'
      FYI='Heptane, C_7 H_16'
      MW_FUEL=100.
      NU_O2=11.
      NU_CO2=7.
      NU_H2O=8.
      CO_YIELD=0.006
      SOOT_YIELD=0.015 /

*** SPECIFY BOUNDARY CONDITIONS OF MESHES

ALLOW 'FLOOR' OF PLUME MESH TO REMAIN INERT BY DEFAULT

&VENT XB=10.25, 10.25, 10.25, 20.25, 0, 6.1, SURF_ID='OPEN'/LEFT
SIDE OF PLUME MESH
&VENT XB=20.25, 20.25, 10.25, 20.25, 0, 6.1,
SURF_ID='OPEN'/RIGHT SIDE OF PLUME MESH
&VENT XB=10.25, 20.25, 10.25, 10.25, 0, 6.1,
SURF_ID='OPEN'/BOTTOM (YBAR0) OF PLUME MESH
&VENT XB=10.25, 20.25, 10.25, 20.25, 6.1, 6.1, SURF_ID='CEILING
TILE'/CEILING OF PLUME MESH
&VENT XB=10.25, 20.25, 20.25, 20.25, 0, 6.1, SURF_ID='OPEN'/TOP
(YBAR) OF PLUME MESH


&VENT XB=10.25, 20.25, 20.25, 20.25, 3.05, 6.1,
SURF_ID='OPEN'/BOTTOM (YBAR0) OF CEILING JET MESH
&VENT XB=10.25, 10.25, 20.25, 30.25, 3.05, 6.1,
SURF_ID='OPEN'/LEFT SIDE OF CEILING JET MESH
&VENT XB=20.25, 20.25, 20.25, 30.25, 3.05, 6.1,
SURF_ID='OPEN'/RIGHT SIDE OF CEILING JET MESH
```

```
&VENT XB=10.25, 20.25, 30.25, 30.25, 3.05, 6.1,
SURF_ID='OPEN'/TOP (YBAR) OF CEILING JET MESH
&VENT XB=10.25, 20.25, 20.25, 30.25, 3.05, 3.05,
SURF_ID='OPEN'/FLOOR OF CEILING JET MESH
&VENT XB=10.25, 20.25, 20.25, 30.25, 6.1, 6.1, SURF_ID='CEILING
TILE'/CEILING OF CEILING JET MESH

&OBST XB= 14.75,15.75,14.75,15.75,0,0.6, RGB=0,0,1/ burner 1 m x
1 m (39.37 inches by 39.37 inches)

&VENT XB= 14.75,15.75,14.75,15.75,0.6,0.6,VENT_COLOR='RED',
SURF_ID='BURN'/ fire

&SURF ID='BURN',HRRPUA=7500, RAMP_Q='FIRE'/

&RAMP ID='FIRE', T=0, F=0.0025/
&RAMP ID='FIRE', T=5, F=0.0056/
&RAMP ID='FIRE', T=10, F=0.01/
&RAMP ID='FIRE', T=15, F=0.0156/
&RAMP ID='FIRE', T=20, F=0.0225/
&RAMP ID='FIRE', T=25, F=0.0306/
&RAMP ID='FIRE', T=30, F=0.04/
&RAMP ID='FIRE', T=35, F=0.0506/
&RAMP ID='FIRE', T=40, F=0.0624/
&RAMP ID='FIRE', T=45, F=0.0656/
&RAMP ID='FIRE', T=50, F=0.0688/
&RAMP ID='FIRE', T=55, F=0.0721/
&RAMP ID='FIRE', T=60, F=0.0755/
&RAMP ID='FIRE', T=65, F=0.079/
&RAMP ID='FIRE', T=70, F=0.0825/
&RAMP ID='FIRE', T=75, F=0.0862/
&RAMP ID='FIRE', T=80, F=0.0899/
&RAMP ID='FIRE', T=85, F=0.0936/
&RAMP ID='FIRE', T=90, F=0.0975/
&RAMP ID='FIRE', T=95, F=0.1014/
&RAMP ID='FIRE', T=100, F=0.1055/
&RAMP ID='FIRE', T=105, F=0.1096/
&RAMP ID='FIRE', T=110, F=0.1137/
&RAMP ID='FIRE', T=115, F=0.118/
&RAMP ID='FIRE', T=120, F=0.1223/
&RAMP ID='FIRE', T=125, F=0.1267/
&RAMP ID='FIRE', T=130, F=0.1312/
&RAMP ID='FIRE', T=135, F=0.1358/
&RAMP ID='FIRE', T=140, F=0.1404/
&RAMP ID='FIRE', T=145, F=0.1451/
&RAMP ID='FIRE', T=150, F=0.1499/
&RAMP ID='FIRE', T=155, F=0.1548/
&RAMP ID='FIRE', T=160, F=0.1597/
&RAMP ID='FIRE', T=165, F=0.1648/
&RAMP ID='FIRE', T=170, F=0.1699/
&RAMP ID='FIRE', T=175, F=0.1751/
&RAMP ID='FIRE', T=180, F=0.1803/
&RAMP ID='FIRE', T=185, F=0.1857/
&RAMP ID='FIRE', T=190, F=0.1911/
&RAMP ID='FIRE', T=195, F=0.1966/
```

```
&RAMP ID='FIRE', T=200, F=0.2022/
&RAMP ID='FIRE', T=205, F=0.2078/
&RAMP ID='FIRE', T=210, F=0.2136/
&RAMP ID='FIRE', T=215, F=0.2194/
&RAMP ID='FIRE', T=220, F=0.2253/
&RAMP ID='FIRE', T=225, F=0.2312/
&RAMP ID='FIRE', T=230, F=0.2373/
&RAMP ID='FIRE', T=235, F=0.2434/
&RAMP ID='FIRE', T=240, F=0.2496/
&RAMP ID='FIRE', T=245, F=0.2559/
&RAMP ID='FIRE', T=250, F=0.2622/
&RAMP ID='FIRE', T=255, F=0.2687/
&RAMP ID='FIRE', T=260, F=0.2752/
&RAMP ID='FIRE', T=265, F=0.2818/
&RAMP ID='FIRE', T=270, F=0.2884/
&RAMP ID='FIRE', T=275, F=0.2952/
&RAMP ID='FIRE', T=280, F=0.302/
&RAMP ID='FIRE', T=285, F=0.3089/
&RAMP ID='FIRE', T=290, F=0.3159/
&RAMP ID='FIRE', T=295, F=0.323/
&RAMP ID='FIRE', T=300, F=0.3301/
&RAMP ID='FIRE', T=305, F=0.3373/
&RAMP ID='FIRE', T=310, F=0.3446/
&RAMP ID='FIRE', T=315, F=0.352/
&RAMP ID='FIRE', T=320, F=0.3594/
&RAMP ID='FIRE', T=325, F=0.367/
&RAMP ID='FIRE', T=330, F=0.3746/
&RAMP ID='FIRE', T=335, F=0.3822/
&RAMP ID='FIRE', T=340, F=0.39/
&RAMP ID='FIRE', T=345, F=0.3978/
&RAMP ID='FIRE', T=350, F=0.4058/
&RAMP ID='FIRE', T=355, F=0.4138/
&RAMP ID='FIRE', T=360, F=0.4218/
&RAMP ID='FIRE', T=365, F=0.43/
&RAMP ID='FIRE', T=370, F=0.4382/
&RAMP ID='FIRE', T=375, F=0.4465/
&RAMP ID='FIRE', T=380, F=0.4549/
&RAMP ID='FIRE', T=385, F=0.4634/
&RAMP ID='FIRE', T=390, F=0.4719/
&RAMP ID='FIRE', T=395, F=0.4805/
&RAMP ID='FIRE', T=400, F=0.4892/
&RAMP ID='FIRE', T=405, F=0.498/
&RAMP ID='FIRE', T=410, F=0.5068/
&RAMP ID='FIRE', T=415, F=0.5158/
&RAMP ID='FIRE', T=420, F=0.5248/
&RAMP ID='FIRE', T=425, F=0.5339/
&RAMP ID='FIRE', T=430, F=0.543/
&RAMP ID='FIRE', T=435, F=0.5523/
&RAMP ID='FIRE', T=440, F=0.5616/
&RAMP ID='FIRE', T=445, F=0.571/
&RAMP ID='FIRE', T=450, F=0.5805/
&RAMP ID='FIRE', T=455, F=0.59/
&RAMP ID='FIRE', T=460, F=0.5997/
&RAMP ID='FIRE', T=465, F=0.6094/
&RAMP ID='FIRE', T=470, F=0.6192/
```

```
&RAMP ID='FIRE', T=475, F=0.629/
&RAMP ID='FIRE', T=480, F=0.639/
&RAMP ID='FIRE', T=485, F=0.649/
&RAMP ID='FIRE', T=490, F=0.6591/
&RAMP ID='FIRE', T=495, F=0.6693/
&RAMP ID='FIRE', T=500, F=0.6795/
&RAMP ID='FIRE', T=505, F=0.6899/
&RAMP ID='FIRE', T=510, F=0.7003/
&RAMP ID='FIRE', T=515, F=0.7108/
&RAMP ID='FIRE', T=520, F=0.7213/
&RAMP ID='FIRE', T=525, F=0.732/
&RAMP ID='FIRE', T=530, F=0.7427/
&RAMP ID='FIRE', T=535, F=0.7535/
&RAMP ID='FIRE', T=540, F=0.7644/
&RAMP ID='FIRE', T=545, F=0.7754/
&RAMP ID='FIRE', T=550, F=0.7864/
&RAMP ID='FIRE', T=555, F=0.7975/
&RAMP ID='FIRE', T=560, F=0.8087/
&RAMP ID='FIRE', T=565, F=0.82/
&RAMP ID='FIRE', T=570, F=0.8313/
&RAMP ID='FIRE', T=575, F=0.8428/
&RAMP ID='FIRE', T=580, F=0.8543/
&RAMP ID='FIRE', T=585, F=0.8658/
&RAMP ID='FIRE', T=590, F=0.8775/
&RAMP ID='FIRE', T=595, F=0.8892/
&RAMP ID='FIRE', T=600, F=0.9011/

&THCP XYZ= 15.25, 15.25, 6, QUANTITY='TEMPERATURE',
LABEL='PLUME' /
&HEAT XYZ= 15.25, 15.25, 6, RTI=32,
ACTIVATION_TEMPERATURE=1000.0, LABEL='PLUME32'/
&HEAT XYZ= 15.25, 15.25, 6, RTI=164,
ACTIVATION_TEMPERATURE=1000, LABEL='PLUME164' /
&HEAT XYZ= 15.25, 15.25, 6, RTI=287,
ACTIVATION_TEMPERATURE=1000.0, LABEL='PLUME287' /

&THCP XYZ= 15.25, 17.41, 6, QUANTITY='TEMPERATURE', LABEL='7FT'
/
&HEAT XYZ= 15.25, 17.41, 6, RTI=32,
ACTIVATION_TEMPERATURE=1000.0, LABEL='7FT32'/
&HEAT XYZ= 15.25, 17.41, 6, RTI=164,
ACTIVATION_TEMPERATURE=1000, LABEL='7FT164' /
&HEAT XYZ= 15.25, 17.41, 6, RTI=287,
ACTIVATION_TEMPERATURE=1000.0, LABEL='7FT287' /

&THCP XYZ= 15.25, 21.71, 6, QUANTITY='TEMPERATURE', LABEL='21FT'
/
&HEAT XYZ= 15.25, 21.71, 6, RTI=32,
ACTIVATION_TEMPERATURE=1000.0, LABEL='21FT32'/
&HEAT XYZ= 15.25, 21.71, 6, RTI=164,
ACTIVATION_TEMPERATURE=1000, LABEL='21FT164' /
&HEAT XYZ= 15.25, 21.71, 6, RTI=287,
ACTIVATION_TEMPERATURE=1000.0, LABEL='21FT287' /
```

```
&THCP XYZ= 15.25, 26.03, 6, QUANTITY='TEMPERATURE', LABEL='35FT'
/
&HEAT XYZ= 15.25, 26.03, 6, RTI=32,
ACTIVATION_TEMPERATURE=1000.0, LABEL='35FT32'/
&HEAT XYZ= 15.25, 26.03, 6, RTI=164,
ACTIVATION_TEMPERATURE=1000, LABEL='35FT164' /
&HEAT XYZ= 15.25, 26.03, 6, RTI=287,
ACTIVATION_TEMPERATURE=1000.0, LABEL='35FT287' /

&SLCF PBX=15.25, QUANTITY='TEMPERATURE'/
&SLCF PBX=15.25, QUANTITY='VELOCITY'/
```

```
&HEAD CHID='UL25-100mm',TITLE='25 ft ceiling with 100 mm grid
spacing' /


&GRID IBAR=100,JBAR=100,KBAR=80 /
&PDIM XBAR0=10.25, XBAR=20.25, YBAR0=10.25, YBAR=20.25, ZBAR=7.6
/ Plume mesh

&GRID IBAR=100, JBAR=100, KBAR=40/
&PDIM XBAR0=10.25, XBAR=20.25, YBAR0=20.25, YBAR=30.25,
ZBAR0=3.8, ZBAR=7.6/ Ceiling jet mesh

&TIME TWFIN=600 /

&MISC REACTION='HEPTANE', DTCORE=15/

&SURF ID                  = 'CEILING TILE'
      RGB                 = .95,.95,.95
      FYI                 = 'Data as provided by UL'
      KS                  = 0.0611
      DENSITY      = 313
      C_P        = 0.753
      DELTA               = 0.0158
      BACKING             = 'EXPOSED'/

&REAC ID='HEPTANE'
      FYI='Heptane, C_7 H_16'
      MW_FUEL=100.
      NU_O2=11.
      NU_CO2=7.
      NU_H2O=8.
      CO_YIELD=0.006
      SOOT_YIELD=0.015 /

*** SPECIFY BOUNDARY CONDITIONS OF MESHES

ALLOW 'FLOOR' OF PLUME MESH TO REMAIN INERT BY DEFAULT

&VENT XB=10.25, 10.25, 10.25, 20.25, 0, 7.6, SURF_ID='OPEN'/LEFT
SIDE OF PLUME MESH
&VENT XB=20.25, 20.25, 10.25, 20.25, 0, 7.6,
SURF_ID='OPEN'/RIGHT SIDE OF PLUME MESH
&VENT XB=10.25, 20.25, 10.25, 10.25, 0, 7.6,
SURF_ID='OPEN'/BOTTOM (YBAR0) OF PLUME MESH
&VENT XB=10.25, 20.25, 10.25, 20.25, 7.6, 7.6, SURF_ID='CEILING
TILE'/CEILING OF PLUME MESH
&VENT XB=10.25, 20.25, 20.25, 20.25, 0, 7.6, SURF_ID='OPEN'/TOP
(YBAR) OF PLUME MESH

&VENT XB=10.25, 20.25, 20.25, 20.25, 3.8, 7.6,
SURF_ID='OPEN'/BOTTOM (YBAR0) OF CEILING JET MESH
&VENT XB=10.25, 10.25, 20.25, 30.25, 3.8, 7.6,
SURF_ID='OPEN'/LEFT SIDE OF CEILING JET MESH
&VENT XB=20.25, 20.25, 20.25, 30.25, 3.8, 7.6,
SURF_ID='OPEN'/RIGHT SIDE OF CEILING JET MESH
```

```
&VENT XB=10.25, 20.25, 30.25, 30.25, 3.8, 7.6,
SURF_ID='OPEN'/TOP (YBAR) OF CEILING JET MESH
&VENT XB=10.25, 20.25, 20.25, 30.25, 3.8, 3.8,
SURF_ID='OPEN'/FLOOR OF CEILING JET MESH
&VENT XB=10.25, 20.25, 20.25, 30.25, 7.6, 7.6, SURF_ID='CEILING
TILE'/CEILING OF CEILING JET MESH

&OBST XB= 14.75,15.75,14.75,15.75,0,0.6, RGB=0,0,1/ burner 1 m x
1 m (39.37 inches by 39.37 inches)

&VENT XB= 14.75,15.75,14.75,15.75,0.6,0.6,VENT_COLOR='RED',
SURF_ID='BURN'/ fire

&SURF ID='BURN',HRRPUA=7500, RAMP_Q='FIRE'/

&RAMP ID='FIRE', T=0, F=0.0025/
&RAMP ID='FIRE', T=5, F=0.0056/
&RAMP ID='FIRE', T=10, F=0.01/
&RAMP ID='FIRE', T=15, F=0.0156/
&RAMP ID='FIRE', T=20, F=0.0225/
&RAMP ID='FIRE', T=25, F=0.0306/
&RAMP ID='FIRE', T=30, F=0.04/
&RAMP ID='FIRE', T=35, F=0.0506/
&RAMP ID='FIRE', T=40, F=0.0624/
&RAMP ID='FIRE', T=45, F=0.0656/
&RAMP ID='FIRE', T=50, F=0.0688/
&RAMP ID='FIRE', T=55, F=0.0721/
&RAMP ID='FIRE', T=60, F=0.0755/
&RAMP ID='FIRE', T=65, F=0.079/
&RAMP ID='FIRE', T=70, F=0.0825/
&RAMP ID='FIRE', T=75, F=0.0862/
&RAMP ID='FIRE', T=80, F=0.0899/
&RAMP ID='FIRE', T=85, F=0.0936/
&RAMP ID='FIRE', T=90, F=0.0975/
&RAMP ID='FIRE', T=95, F=0.1014/
&RAMP ID='FIRE', T=100, F=0.1055/
&RAMP ID='FIRE', T=105, F=0.1096/
&RAMP ID='FIRE', T=110, F=0.1137/
&RAMP ID='FIRE', T=115, F=0.118/
&RAMP ID='FIRE', T=120, F=0.1223/
&RAMP ID='FIRE', T=125, F=0.1267/
&RAMP ID='FIRE', T=130, F=0.1312/
&RAMP ID='FIRE', T=135, F=0.1358/
&RAMP ID='FIRE', T=140, F=0.1404/
&RAMP ID='FIRE', T=145, F=0.1451/
&RAMP ID='FIRE', T=150, F=0.1499/
&RAMP ID='FIRE', T=155, F=0.1548/
&RAMP ID='FIRE', T=160, F=0.1597/
&RAMP ID='FIRE', T=165, F=0.1648/
&RAMP ID='FIRE', T=170, F=0.1699/
&RAMP ID='FIRE', T=175, F=0.1751/
&RAMP ID='FIRE', T=180, F=0.1803/
&RAMP ID='FIRE', T=185, F=0.1857/
&RAMP ID='FIRE', T=190, F=0.1911/
&RAMP ID='FIRE', T=195, F=0.1966/
```

```
&RAMP ID='FIRE',  T=200,  F=0.2022/
&RAMP ID='FIRE',  T=205,  F=0.2078/
&RAMP ID='FIRE',  T=210,  F=0.2136/
&RAMP ID='FIRE',  T=215,  F=0.2194/
&RAMP ID='FIRE',  T=220,  F=0.2253/
&RAMP ID='FIRE',  T=225,  F=0.2312/
&RAMP ID='FIRE',  T=230,  F=0.2373/
&RAMP ID='FIRE',  T=235,  F=0.2434/
&RAMP ID='FIRE',  T=240,  F=0.2496/
&RAMP ID='FIRE',  T=245,  F=0.2559/
&RAMP ID='FIRE',  T=250,  F=0.2622/
&RAMP ID='FIRE',  T=255,  F=0.2687/
&RAMP ID='FIRE',  T=260,  F=0.2752/
&RAMP ID='FIRE',  T=265,  F=0.2818/
&RAMP ID='FIRE',  T=270,  F=0.2884/
&RAMP ID='FIRE',  T=275,  F=0.2952/
&RAMP ID='FIRE',  T=280,  F=0.302/
&RAMP ID='FIRE',  T=285,  F=0.3089/
&RAMP ID='FIRE',  T=290,  F=0.3159/
&RAMP ID='FIRE',  T=295,  F=0.323/
&RAMP ID='FIRE',  T=300,  F=0.3301/
&RAMP ID='FIRE',  T=305,  F=0.3373/
&RAMP ID='FIRE',  T=310,  F=0.3446/
&RAMP ID='FIRE',  T=315,  F=0.352/
&RAMP ID='FIRE',  T=320,  F=0.3594/
&RAMP ID='FIRE',  T=325,  F=0.367/
&RAMP ID='FIRE',  T=330,  F=0.3746/
&RAMP ID='FIRE',  T=335,  F=0.3822/
&RAMP ID='FIRE',  T=340,  F=0.39/
&RAMP ID='FIRE',  T=345,  F=0.3978/
&RAMP ID='FIRE',  T=350,  F=0.4058/
&RAMP ID='FIRE',  T=355,  F=0.4138/
&RAMP ID='FIRE',  T=360,  F=0.4218/
&RAMP ID='FIRE',  T=365,  F=0.43/
&RAMP ID='FIRE',  T=370,  F=0.4382/
&RAMP ID='FIRE',  T=375,  F=0.4465/
&RAMP ID='FIRE',  T=380,  F=0.4549/
&RAMP ID='FIRE',  T=385,  F=0.4634/
&RAMP ID='FIRE',  T=390,  F=0.4719/
&RAMP ID='FIRE',  T-395,  F-0.4805/
&RAMP ID='FIRE',  T=400,  F=0.4892/
&RAMP ID='FIRE',  T=405,  F=0.498/
&RAMP ID='FIRE',  T=410,  F=0.5068/
&RAMP ID='FIRE',  T=415,  F=0.5158/
&RAMP ID='FIRE',  T=420,  F=0.5248/
&RAMP ID='FIRE',  T=425,  F=0.5339/
&RAMP ID='FIRE',  T=430,  F=0.543/
&RAMP ID='FIRE',  T=435,  F=0.5523/
&RAMP ID='FIRE',  T=440,  F=0.5616/
&RAMP ID='FIRE',  T=445,  F=0.571/
&RAMP ID='FIRE',  T=450,  F=0.5805/
&RAMP ID='FIRE',  T=455,  F=0.59/
&RAMP ID='FIRE',  T=460,  F=0.5997/
&RAMP ID='FIRE',  T=465,  F=0.6094/
&RAMP ID='FIRE',  T=470,  F=0.6192/
```

```
&RAMP ID='FIRE', T=475, F=0.629/
&RAMP ID='FIRE', T=480, F=0.639/
&RAMP ID='FIRE', T=485, F=0.649/
&RAMP ID='FIRE', T=490, F=0.6591/
&RAMP ID='FIRE', T=495, F=0.6693/
&RAMP ID='FIRE', T=500, F=0.6795/
&RAMP ID='FIRE', T=505, F=0.6899/
&RAMP ID='FIRE', T=510, F=0.7003/
&RAMP ID='FIRE', T=515, F=0.7108/
&RAMP ID='FIRE', T=520, F=0.7213/
&RAMP ID='FIRE', T=525, F=0.732/
&RAMP ID='FIRE', T=530, F=0.7427/
&RAMP ID='FIRE', T=535, F=0.7535/
&RAMP ID='FIRE', T=540, F=0.7644/
&RAMP ID='FIRE', T=545, F=0.7754/
&RAMP ID='FIRE', T=550, F=0.7864/
&RAMP ID='FIRE', T=555, F=0.7975/
&RAMP ID='FIRE', T=560, F=0.8087/
&RAMP ID='FIRE', T=565, F=0.82/
&RAMP ID='FIRE', T=570, F=0.8313/
&RAMP ID='FIRE', T=575, F=0.8428/
&RAMP ID='FIRE', T=580, F=0.8543/
&RAMP ID='FIRE', T=585, F=0.8658/
&RAMP ID='FIRE', T=590, F=0.8775/
&RAMP ID='FIRE', T=595, F=0.8892/
&RAMP ID='FIRE', T=600, F=0.9011/

&THCP XYZ= 15.25, 15.25, 7.5, QUANTITY='TEMPERATURE',
LABEL='PLUME' /
&HEAT XYZ= 15.25, 15.25, 7.5, RTI=32,
ACTIVATION_TEMPERATURE=1000.0, LABEL='PLUME32'/
&HEAT XYZ= 15.25, 15.25, 7.5, RTI=164,
ACTIVATION_TEMPERATURE=1000, LABEL='PLUME164' /
&HEAT XYZ= 15.25, 15.25, 7.5, RTI=287,
ACTIVATION_TEMPERATURE=1000.0, LABEL='PLUME287' /

&THCP XYZ= 15.25, 17.41, 7.5, QUANTITY='TEMPERATURE',
LABEL='7FT' /
&HEAT XYZ= 15.25, 17.41, 7.5, RTI=32,
ACTIVATION_TEMPERATURE=1000.0, LABEL='7FT32'/
&HEAT XYZ= 15.25, 17.41, 7.5, RTI=164,
ACTIVATION_TEMPERATURE=1000, LABEL='7FT164' /
&HEAT XYZ= 15.25, 17.41, 7.5, RTI=287,
ACTIVATION_TEMPERATURE=1000.0, LABEL='7FT287' /

&THCP XYZ= 15.25, 21.71, 7.5, QUANTITY='TEMPERATURE',
LABEL='21FT' /
&HEAT XYZ= 15.25, 21.71, 7.5, RTI=32,
ACTIVATION_TEMPERATURE=1000.0, LABEL='21FT32'/
&HEAT XYZ= 15.25, 21.71, 7.5, RTI=164,
ACTIVATION_TEMPERATURE=1000, LABEL='21FT164' /
&HEAT XYZ= 15.25, 21.71, 7.5, RTI=287,
ACTIVATION_TEMPERATURE=1000.0, LABEL='21FT287' /
```

```
&THCP XYZ= 15.25, 26.03, 7.5, QUANTITY='TEMPERATURE',
LABEL='35FT' /
&HEAT XYZ= 15.25, 26.03, 7.5, RTI=32,
ACTIVATION_TEMPERATURE=1000.0, LABEL='35FT32'/
&HEAT XYZ= 15.25, 26.03, 7.5, RTI=164,
ACTIVATION_TEMPERATURE=1000, LABEL='35FT164' /
&HEAT XYZ= 15.25, 26.03, 7.5, RTI=287,
ACTIVATION_TEMPERATURE=1000.0, LABEL='35FT287' /

&SLCF PBX=15.25, QUANTITY='TEMPERATURE'/
&SLCF PBX=15.25, QUANTITY='VELOCITY'/
```

```
&HEAD CHID='UL35-100mm',TITLE='35 ft ceiling with 100 mm grid
spacing' /


&GRID IBAR=100,JBAR=100,KBAR=108 /
&PDIM XBAR0=10.25, XBAR=20.25, YBAR0=10.25, YBAR=20.25,
ZBAR=10.7 / Plume mesh

&GRID IBAR=100, JBAR=100, KBAR=54/
&PDIM XBAR0=10.25, XBAR=20.25, YBAR0=20.25, YBAR=30.25,
ZBAR0=5.35, ZBAR=10.7/ Ceiling jet mesh

&TIME TWFIN=300 /

&MISC REACTION='HEPTANE', DTCORE=15/

&SURF ID                  = 'CEILING TILE'
      RGB                 = .95,.95,.95
      FYI                 = 'Data as provided by UL'
      KS                  = 0.0611
      DENSITY       = 313
      C_P       = 0.753
      DELTA               = 0.0158
      BACKING             = 'EXPOSED'/

&REAC ID='HEPTANE'
      FYI='Heptane, C_7 H_16'
      MW_FUEL=100.
      NU_O2=11.
      NU_CO2=7.
      NU_H2O=8.
      CO_YIELD=0.006
      SOOT_YIELD=0.015 /

*** SPECIFY BOUNDARY CONDITIONS OF MESHES

ALLOW 'FLOOR' OF PLUME MESH TO REMAIN INERT BY DEFAULT

&VENT XB=10.25, 10.25, 10.25, 20.25, 0, 10.7,
SURF_ID='OPEN'/LEFT SIDE OF PLUME MESH
&VENT XB=20.25, 20.25, 10.25, 20.25, 0, 10.7,
SURF_ID='OPEN'/RIGHT SIDE OF PLUME MESH
&VENT XB=10.25, 20.25, 10.25, 10.25, 0, 10.7,
SURF_ID='OPEN'/BOTTOM (YBAR0) OF PLUME MESH
&VENT XB=10.25, 20.25, 10.25, 20.25, 10.7, 10.7,
SURF_ID='CEILING TILE'/CEILING OF PLUME MESH
&VENT XB=10.25, 20.25, 20.25, 20.25, 0, 10.7, SURF_ID='OPEN'/TOP
(YBAR) OF PLUME MESH

&VENT XB=10.25, 20.25, 20.25, 20.25, 5.35, 10.7,
SURF_ID='OPEN'/BOTTOM (YBAR0) OF CEILING JET MESH
&VENT XB=10.25, 10.25, 20.25, 30.25, 5.35, 10.7,
SURF_ID='OPEN'/LEFT SIDE OF CEILING JET MESH
&VENT XB=20.25, 20.25, 20.25, 30.25, 5.35, 10.7,
SURF_ID='OPEN'/RIGHT SIDE OF CEILING JET MESH
```

```
&VENT XB=10.25, 20.25, 30.25, 30.25, 5.35, 10.7,
SURF_ID='OPEN'/TOP (YBAR) OF CEILING JET MESH
&VENT XB=10.25, 20.25, 20.25, 30.25, 5.35, 5.35,
SURF_ID='OPEN'/FLOOR OF CEILING JET MESH
&VENT XB=10.25, 20.25, 20.25, 30.25, 10.7, 10.7,
SURF_ID='CEILING TILE'/CEILING OF CEILING JET MESH

&OBST XB= 14.75,15.75,14.75,15.75,0,0.6, RGB=0,0,1/ burner 1 m x
1 m (39.37 inches by 39.37 inches)

&VENT XB= 14.75,15.75,14.75,15.75,0.6,0.6,VENT_COLOR='RED',
SURF_ID='BURN'/ fire

&SURF ID='BURN',HRRPUA=10000, RAMP_Q='FIRE'/

&RAMP ID='FIRE', T=0, F=0.0019/
&RAMP ID='FIRE', T=5, F=0.0042/
&RAMP ID='FIRE', T=10, F=0.0075/
&RAMP ID='FIRE', T=15, F=0.0117/
&RAMP ID='FIRE', T=20, F=0.0169/
&RAMP ID='FIRE', T=25, F=0.023/
&RAMP ID='FIRE', T=30, F=0.03/
&RAMP ID='FIRE', T=35, F=0.038/
&RAMP ID='FIRE', T=40, F=0.0468/
&RAMP ID='FIRE', T=45, F=0.0492/
&RAMP ID='FIRE', T=50, F=0.0516/
&RAMP ID='FIRE', T=55, F=0.0541/
&RAMP ID='FIRE', T=60, F=0.0566/
&RAMP ID='FIRE', T=65, F=0.0592/
&RAMP ID='FIRE', T=70, F=0.0619/
&RAMP ID='FIRE', T=75, F=0.0646/
&RAMP ID='FIRE', T=80, F=0.0674/
&RAMP ID='FIRE', T=85, F=0.0702/
&RAMP ID='FIRE', T=90, F=0.0731/
&RAMP ID='FIRE', T=95, F=0.0761/
&RAMP ID='FIRE', T=100, F=0.0791/
&RAMP ID='FIRE', T=105, F=0.0822/
&RAMP ID='FIRE', T=110, F=0.0853/
&RAMP ID='FIRE', T=115, F=0.0885/
&RAMP ID='FIRE', T-120, F-0.0917/
&RAMP ID='FIRE', T=125, F=0.095/
&RAMP ID='FIRE', T=130, F=0.0984/
&RAMP ID='FIRE', T=135, F=0.1018/
&RAMP ID='FIRE', T=140, F=0.1053/
&RAMP ID='FIRE', T=145, F=0.1088/
&RAMP ID='FIRE', T=150, F=0.1124/
&RAMP ID='FIRE', T=155, F=0.1161/
&RAMP ID='FIRE', T=160, F=0.1198/
&RAMP ID='FIRE', T=165, F=0.1236/
&RAMP ID='FIRE', T=170, F=0.1274/
&RAMP ID='FIRE', T=175, F=0.1313/
&RAMP ID='FIRE', T=180, F=0.1353/
&RAMP ID='FIRE', T=185, F=0.1393/
&RAMP ID='FIRE', T=190, F=0.1433/
&RAMP ID='FIRE', T=195, F=0.1474/
```

```
&RAMP ID='FIRE', T=200, F=0.1516/
&RAMP ID='FIRE', T=205, F=0.1559/
&RAMP ID='FIRE', T=210, F=0.1602/
&RAMP ID='FIRE', T=215, F=0.1645/
&RAMP ID='FIRE', T=220, F=0.1689/
&RAMP ID='FIRE', T=225, F=0.1734/
&RAMP ID='FIRE', T=230, F=0.178/
&RAMP ID='FIRE', T=235, F=0.1825/
&RAMP ID='FIRE', T=240, F=0.1872/
&RAMP ID='FIRE', T=245, F=0.1919/
&RAMP ID='FIRE', T=250, F=0.1967/
&RAMP ID='FIRE', T=255, F=0.2015/
&RAMP ID='FIRE', T=260, F=0.2064/
&RAMP ID='FIRE', T=265, F=0.2113/
&RAMP ID='FIRE', T=270, F=0.2163/
&RAMP ID='FIRE', T=275, F=0.2214/
&RAMP ID='FIRE', T=280, F=0.2265/
&RAMP ID='FIRE', T=285, F=0.2317/
&RAMP ID='FIRE', T=290, F=0.2369/
&RAMP ID='FIRE', T=295, F=0.2422/
&RAMP ID='FIRE', T=300, F=0.2476/

&THCP XYZ= 15.25, 15.25, 10.6, QUANTITY='TEMPERATURE',
LABEL='PLUME' /
&HEAT XYZ= 15.25, 15.25, 10.6, RTI=32,
ACTIVATION_TEMPERATURE=1000.0, LABEL='PLUME32'/
&HEAT XYZ= 15.25, 15.25, 10.6, RTI=164,
ACTIVATION_TEMPERATURE=1000, LABEL='PLUME164' /
&HEAT XYZ= 15.25, 15.25, 10.6, RTI=287,
ACTIVATION_TEMPERATURE=1000.0, LABEL='PLUME287' /

&THCP XYZ= 15.25, 17.41, 10.6, QUANTITY='TEMPERATURE',
LABEL='7FT' /
&HEAT XYZ= 15.25, 17.41, 10.6, RTI=32,
ACTIVATION_TEMPERATURE=1000.0, LABEL='7FT32'/
&HEAT XYZ= 15.25, 17.41, 10.6, RTI=164,
ACTIVATION_TEMPERATURE=1000, LABEL='7FT164' /
&HEAT XYZ= 15.25, 17.41, 10.6, RTI=287,
ACTIVATION_TEMPERATURE=1000.0, LABEL='7FT287' /

&THCP XYZ= 15.25, 21.71, 10.6, QUANTITY='TEMPERATURE',
LABEL='21FT' /
&HEAT XYZ= 15.25, 21.71, 10.6, RTI=32,
ACTIVATION_TEMPERATURE=1000.0, LABEL='21FT32'/
&HEAT XYZ= 15.25, 21.71, 10.6, RTI=164,
ACTIVATION_TEMPERATURE=1000, LABEL='21FT164' /
&HEAT XYZ= 15.25, 21.71, 10.6, RTI=287,
ACTIVATION_TEMPERATURE=1000.0, LABEL='21FT287' /

&THCP XYZ= 15.25, 26.03, 10.6, QUANTITY='TEMPERATURE',
LABEL='35FT' /
&HEAT XYZ= 15.25, 26.03, 10.6, RTI=32,
ACTIVATION_TEMPERATURE=1000.0, LABEL='35FT32'/
&HEAT XYZ= 15.25, 26.03, 10.6, RTI=164,
ACTIVATION_TEMPERATURE=1000, LABEL='35FT164' /
```

```
&HEAT XYZ= 15.25, 26.03, 10.6, RTI=287,
ACTIVATION_TEMPERATURE=1000.0, LABEL='35FT287' /

&SLCF PBX=15.25, QUANTITY='TEMPERATURE'/
&SLCF PBX=15.25, QUANTITY='VELOCITY'/
```

```
&HEAD CHID='UL40-100mm',TITLE='40 ft ceiling with 100 mm grid
spacing' /


&GRID IBAR=100,JBAR=100,KBAR=120 /
&PDIM XBAR0=10.25, XBAR=20.25, YBAR0=10.25, YBAR=20.25,
ZBAR=12.2 / Plume mesh

&GRID IBAR=100, JBAR=100, KBAR=60/
&PDIM XBAR0=10.25, XBAR=20.25, YBAR0=20.25, YBAR=30.25,
ZBAR0=6.1, ZBAR=12.2/ Ceiling jet mesh

&TIME TWFIN=300 /

&MISC REACTION='HEPTANE', DTCORE=15/

&SURF ID                  = 'CEILING TILE'
      RGB                 = .95,.95,.95
      FYI                 = 'Data as provided by UL'
      KS                  = 0.0611
      DENSITY      = 313
      C_P       = 0.753
      DELTA               = 0.0158
      BACKING             = 'EXPOSED'/

&REAC ID='HEPTANE'
      FYI='Heptane, C_7 H_16'
      MW_FUEL=100.
      NU_O2=11.
      NU_CO2=7.
      NU_H2O=8.
      CO_YIELD=0.006
      SOOT_YIELD=0.015 /

*** SPECIFY BOUNDARY CONDITIONS OF MESHES

ALLOW 'FLOOR' OF PLUME MESH TO REMAIN INERT BY DEFAULT

&VENT XB=10.25, 10.25, 10.25, 20.25, 0, 12.2,
SURF_ID='OPEN'/LEFT SIDE OF PLUME MESH
&VENT XB=20.25, 20.25, 10.25, 20.25, 0, 12.2,
SURF_ID='OPEN'/RIGHT SIDE OF PLUME MESH
&VENT XB=10.25, 20.25, 10.25, 10.25, 0, 12.2,
SURF_ID='OPEN'/BOTTOM (YBAR0) OF PLUME MESH
&VENT XB=10.25, 20.25, 10.25, 20.25, 12.2, 12.2,
SURF_ID='CEILING TILE'/CEILING OF PLUME MESH
&VENT XB=10.25, 20.25, 20.25, 20.25, 0, 12.2, SURF_ID='OPEN'/TOP
(YBAR) OF PLUME MESH

&VENT XB=10.25, 20.25, 20.25, 20.25, 6.1, 12.2,
SURF_ID='OPEN'/BOTTOM (YBAR0) OF CEILING JET MESH
&VENT XB=10.25, 10.25, 20.25, 30.25, 6.1, 12.2,
SURF_ID='OPEN'/LEFT SIDE OF CEILING JET MESH
&VENT XB=20.25, 20.25, 20.25, 30.25, 6.1, 12.2,
SURF_ID='OPEN'/RIGHT SIDE OF CEILING JET MESH
```

```
&VENT XB=10.25, 20.25, 30.25, 30.25, 6.1, 12.2,
SURF_ID='OPEN'/TOP (YBAR) OF CEILING JET MESH
&VENT XB=10.25, 20.25, 20.25, 30.25, 6.1, 6.1,
SURF_ID='OPEN'/FLOOR OF CEILING JET MESH
&VENT XB=10.25, 20.25, 20.25, 30.25, 12.2, 12.2,
SURF_ID='CEILING TILE'/CEILING OF CEILING JET MESH

&OBST XB= 14.75,15.75,14.75,15.75,0,0.6, RGB=0,0,1/ burner 1 m x
1 m (39.37 inches by 39.37 inches)

&VENT XB= 14.75,15.75,14.75,15.75,0.6,0.6,VENT_COLOR='RED',
SURF_ID='BURN'/ fire

&SURF ID='BURN',HRRPUA=10000, RAMP_Q='FIRE'/

&RAMP ID='FIRE', T=0, F=0.0019/
&RAMP ID='FIRE', T=5, F=0.0042/
&RAMP ID='FIRE', T=10, F=0.0075/
&RAMP ID='FIRE', T=15, F=0.0117/
&RAMP ID='FIRE', T=20, F=0.0169/
&RAMP ID='FIRE', T=25, F=0.023/
&RAMP ID='FIRE', T=30, F=0.03/
&RAMP ID='FIRE', T=35, F=0.038/
&RAMP ID='FIRE', T=40, F=0.0468/
&RAMP ID='FIRE', T=45, F=0.0492/
&RAMP ID='FIRE', T=50, F=0.0516/
&RAMP ID='FIRE', T=55, F=0.0541/
&RAMP ID='FIRE', T=60, F=0.0566/
&RAMP ID='FIRE', T=65, F=0.0592/
&RAMP ID='FIRE', T=70, F=0.0619/
&RAMP ID='FIRE', T=75, F=0.0646/
&RAMP ID='FIRE', T=80, F=0.0674/
&RAMP ID='FIRE', T=85, F=0.0702/
&RAMP ID='FIRE', T=90, F=0.0731/
&RAMP ID='FIRE', T=95, F=0.0761/
&RAMP ID='FIRE', T=100, F=0.0791/
&RAMP ID='FIRE', T=105, F=0.0822/
&RAMP ID='FIRE', T=110, F=0.0853/
&RAMP ID='FIRE', T=115, F=0.0885/
&RAMP ID='FIRE', T=120, F=0.0917/
&RAMP ID='FIRE', T=125, F=0.095/
&RAMP ID='FIRE', T=130, F=0.0984/
&RAMP ID='FIRE', T=135, F=0.1018/
&RAMP ID='FIRE', T=140, F=0.1053/
&RAMP ID='FIRE', T=145, F=0.1088/
&RAMP ID='FIRE', T=150, F=0.1124/
&RAMP ID='FIRE', T=155, F=0.1161/
&RAMP ID='FIRE', T=160, F=0.1198/
&RAMP ID='FIRE', T=165, F=0.1236/
&RAMP ID='FIRE', T=170, F=0.1274/
&RAMP ID='FIRE', T=175, F=0.1313/
&RAMP ID='FIRE', T=180, F=0.1353/
&RAMP ID='FIRE', T=185, F=0.1393/
&RAMP ID='FIRE', T=190, F=0.1433/
&RAMP ID='FIRE', T=195, F=0.1474/
```

```
&RAMP ID='FIRE', T=200, F=0.1516/
&RAMP ID='FIRE', T=205, F=0.1559/
&RAMP ID='FIRE', T=210, F=0.1602/
&RAMP ID='FIRE', T=215, F=0.1645/
&RAMP ID='FIRE', T=220, F=0.1689/
&RAMP ID='FIRE', T=225, F=0.1734/
&RAMP ID='FIRE', T=230, F=0.178/
&RAMP ID='FIRE', T=235, F=0.1825/
&RAMP ID='FIRE', T=240, F=0.1872/
&RAMP ID='FIRE', T=245, F=0.1919/
&RAMP ID='FIRE', T=250, F=0.1967/
&RAMP ID='FIRE', T=255, F=0.2015/
&RAMP ID='FIRE', T=260, F=0.2064/
&RAMP ID='FIRE', T=265, F=0.2113/
&RAMP ID='FIRE', T=270, F=0.2163/
&RAMP ID='FIRE', T=275, F=0.2214/
&RAMP ID='FIRE', T=280, F=0.2265/
&RAMP ID='FIRE', T=285, F=0.2317/
&RAMP ID='FIRE', T=290, F=0.2369/
&RAMP ID='FIRE', T=295, F=0.2422/
&RAMP ID='FIRE', T=300, F=0.2476/

&THCP XYZ= 15.25, 15.25, 12.1, QUANTITY='TEMPERATURE',
LABEL='PLUME' /
&HEAT XYZ= 15.25, 15.25, 12.1, RTI=32,
ACTIVATION_TEMPERATURE=1000.0, LABEL='PLUME32'/
&HEAT XYZ= 15.25, 15.25, 12.1, RTI=164,
ACTIVATION_TEMPERATURE=1000, LABEL='PLUME164' /
&HEAT XYZ= 15.25, 15.25, 12.1, RTI=287,
ACTIVATION_TEMPERATURE=1000.0, LABEL='PLUME287' /

&THCP XYZ= 15.25, 17.41, 12.1, QUANTITY='TEMPERATURE',
LABEL='7FT' /
&HEAT XYZ= 15.25, 17.41, 12.1, RTI=32,
ACTIVATION_TEMPERATURE=1000.0, LABEL='7FT32'/
&HEAT XYZ= 15.25, 17.41, 12.1, RTI=164,
ACTIVATION_TEMPERATURE=1000, LABEL='7FT164' /
&HEAT XYZ= 15.25, 17.41, 12.1, RTI=287,
ACTIVATION_TEMPERATURE=1000.0, LABEL='7FT287' /

&THCP XYZ= 15.25, 21.71, 12.1, QUANTITY='TEMPERATURE',
LABEL='21FT' /
&HEAT XYZ= 15.25, 21.71, 12.1, RTI=32,
ACTIVATION_TEMPERATURE=1000.0, LABEL='21FT32'/
&HEAT XYZ= 15.25, 21.71, 12.1, RTI=164,
ACTIVATION_TEMPERATURE=1000, LABEL='21FT164' /
&HEAT XYZ= 15.25, 21.71, 12.1, RTI=287,
ACTIVATION_TEMPERATURE=1000.0, LABEL='21FT287' /

&THCP XYZ= 15.25, 26.03, 12.1, QUANTITY='TEMPERATURE',
LABEL='35FT' /
&HEAT XYZ= 15.25, 26.03, 12.1, RTI=32,
ACTIVATION_TEMPERATURE=1000.0, LABEL='35FT32'/
&HEAT XYZ= 15.25, 26.03, 12.1, RTI=164,
ACTIVATION_TEMPERATURE=1000, LABEL='35FT164' /
```

```
&HEAT XYZ= 15.25, 26.03, 12.1, RTI=287,
ACTIVATION_TEMPERATURE=1000.0, LABEL='35FT287' /

&SLCF PBX=15.25, QUANTITY='TEMPERATURE'/
          &SLCF PBX=15.25, QUANTITY='VELOCITY'/
```

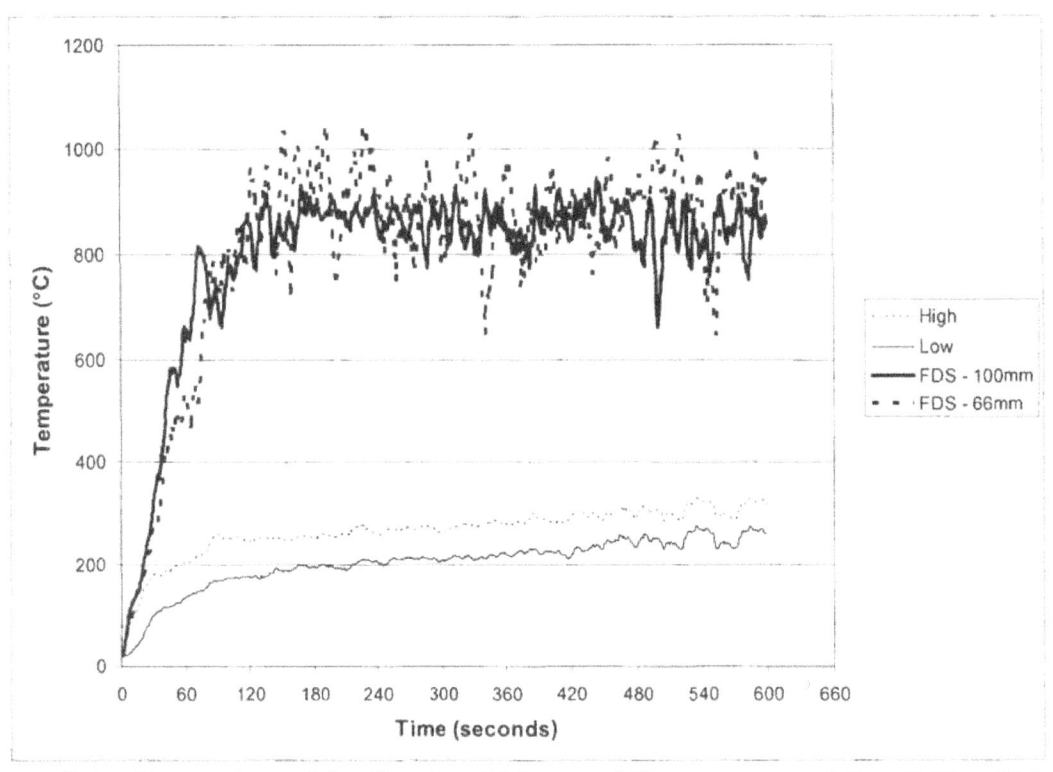

Figure B.1 – Comparison of Predicted and Measured Temperatures, 3.0 m Ceiling Height, Plume Centerline

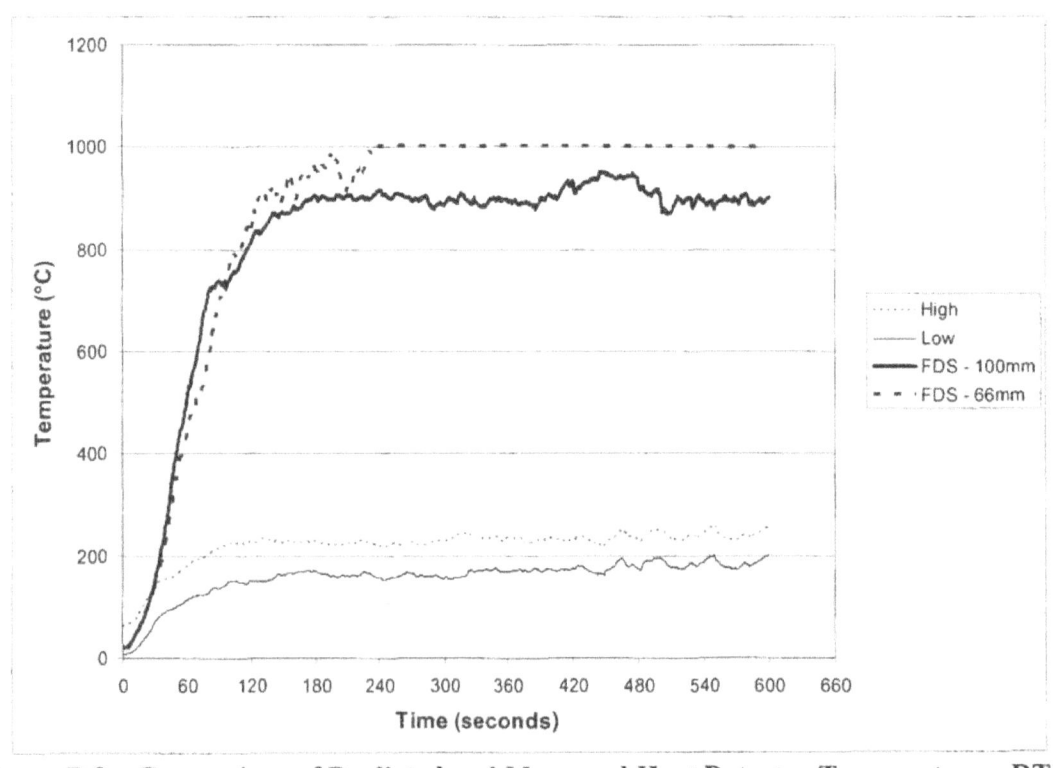

Figure B.2 – Comparison of Predicted and Measured Heat Detector Temperatures, RTI = 32 m$^{1/2}$-s$^{1/2}$, 3.0 m Ceiling Height, Plume Centerline

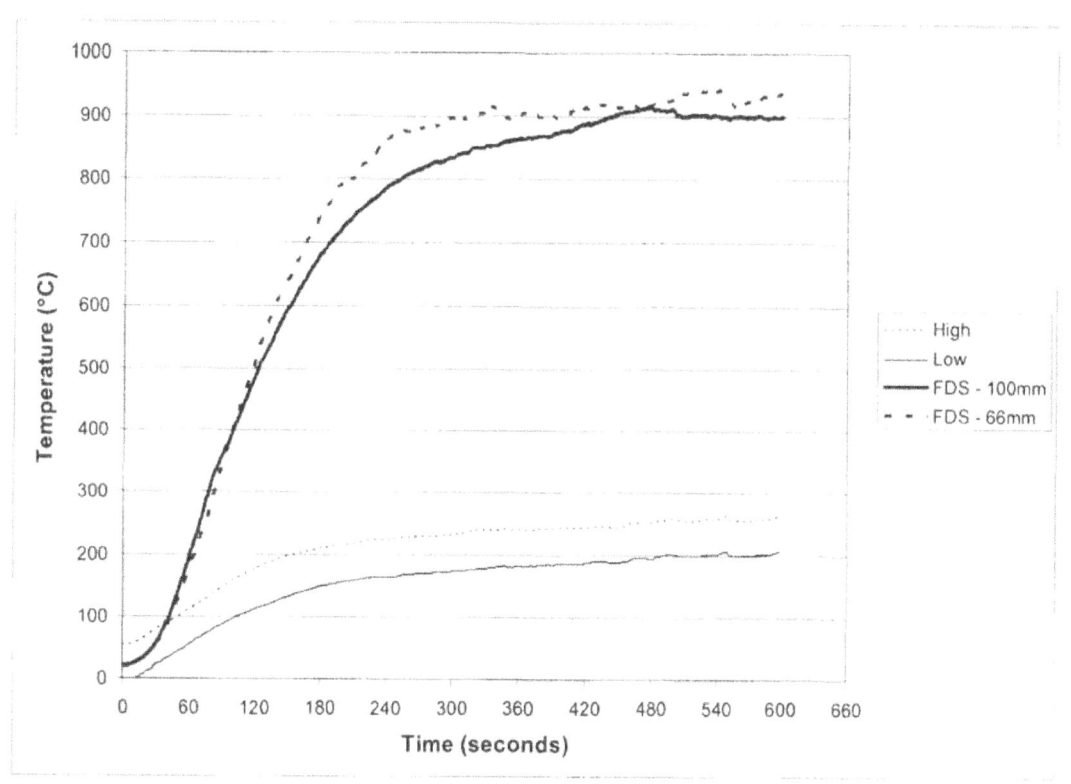

Figure B.3 – Comparison of Predicted and Measured Heat Detector Temperatures, RTI = 164 m$^{1/2}$-s$^{1/2}$, 3.0 m Ceiling Height, Plume Centerline

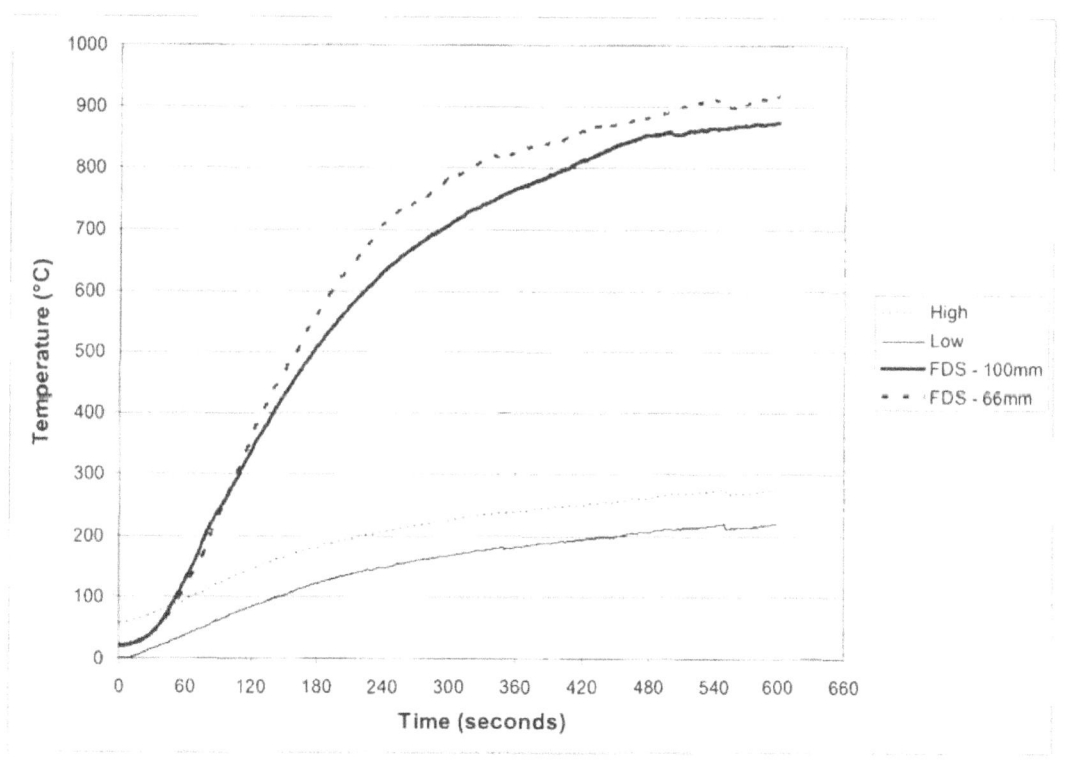

Figure B.4 – Comparison of Predicted and Measured Heat Detector Temperatures, RTI = 287 m$^{1/2}$-s$^{1/2}$, 3.0 m Ceiling Height, Plume Centerline

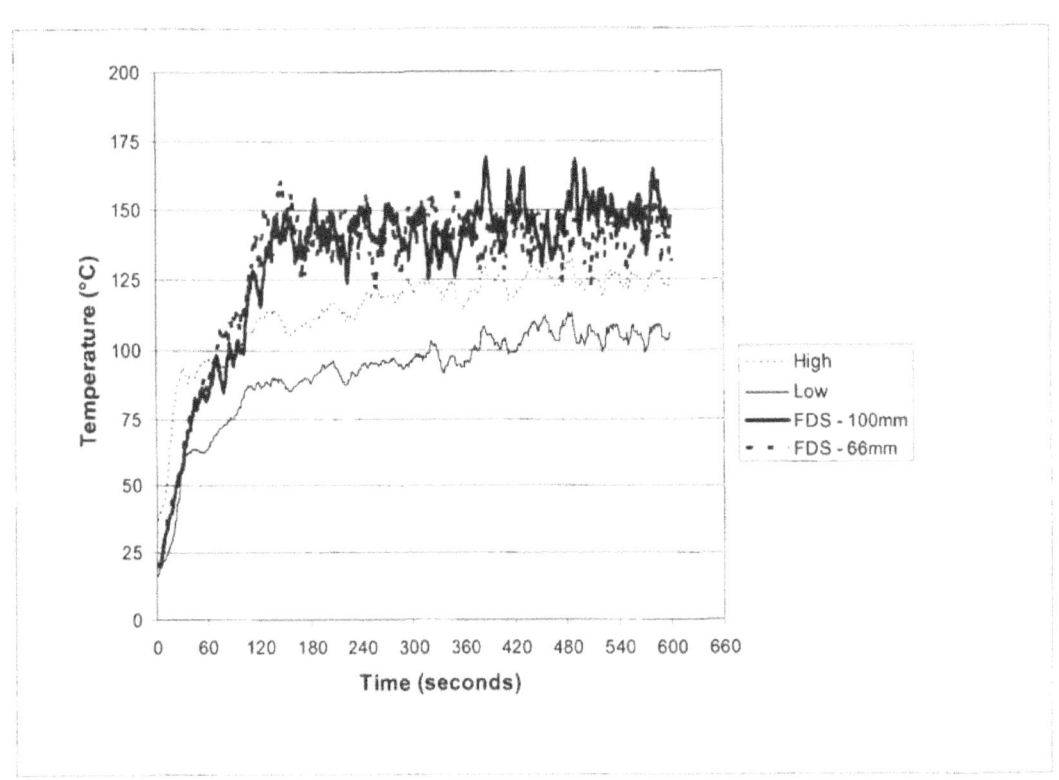

Figure B.5 – Comparison of Predicted and Measured Temperatures, 3.0 m Ceiling Height, Radial Distance = 2.2 m

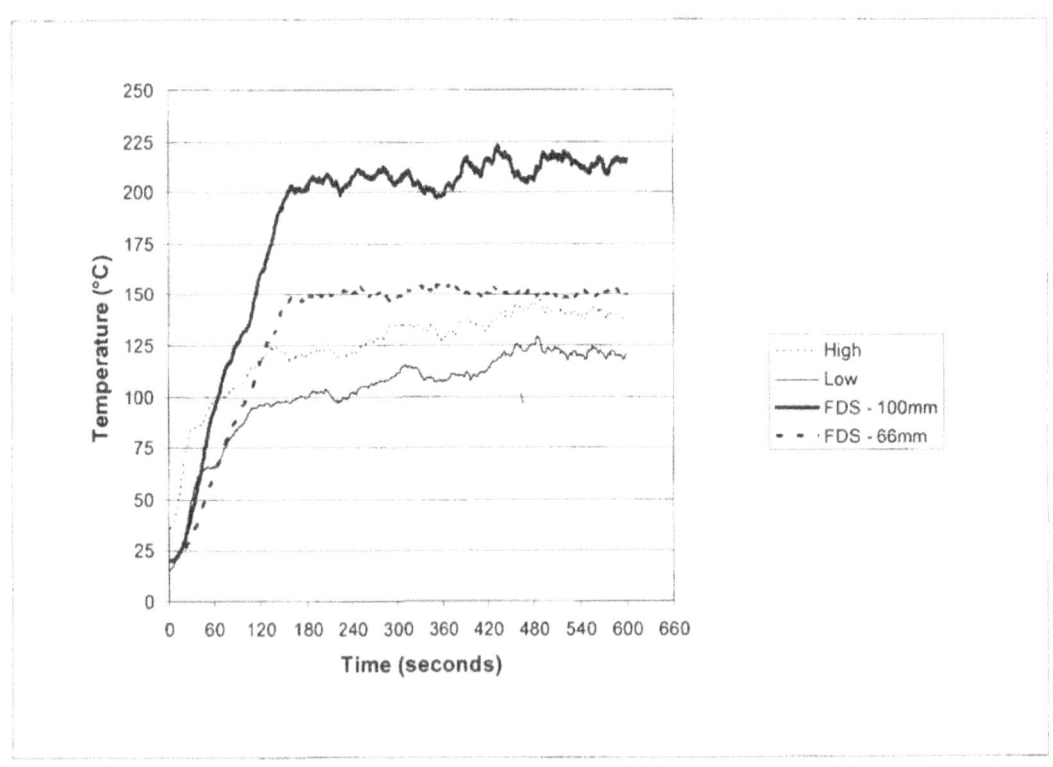

Figure B.6 – Comparison of Predicted and Measured Heat Detector Temperatures, RTI = 32 $m^{1/2}$-$s^{1/2}$, 3.0 m Ceiling Height, Radial Distance = 2.2 m

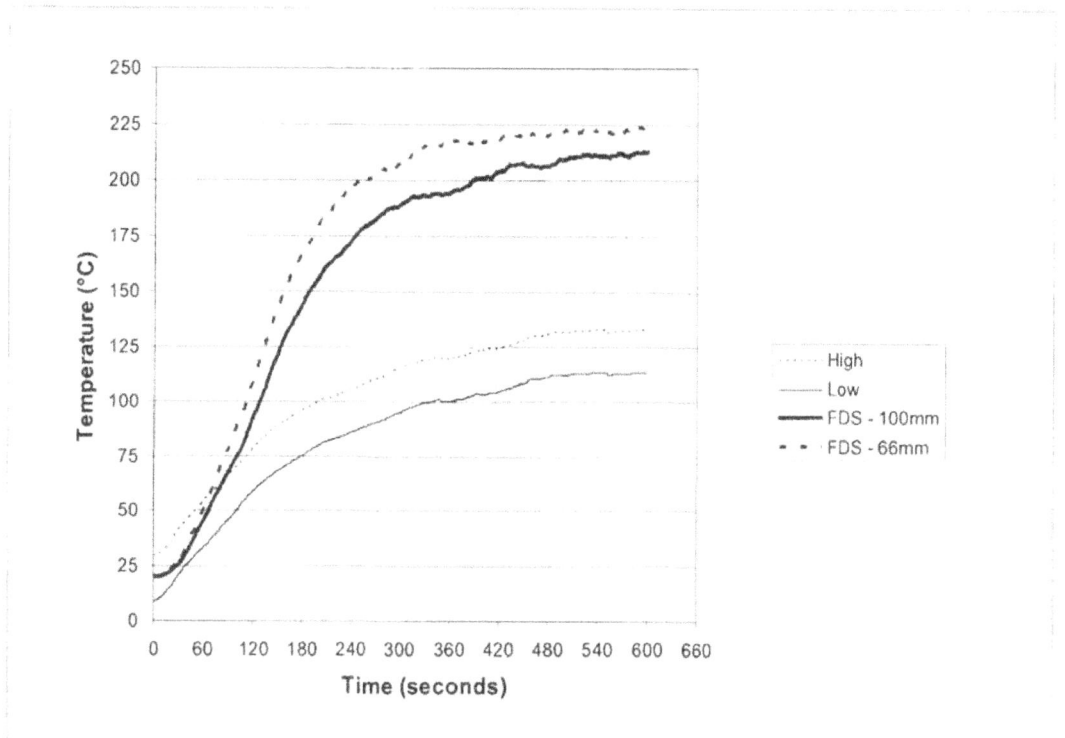

Figure B.7 – Comparison of Predicted and Measured Heat Detector Temperatures, RTI = 164 m$^{1/2}$-s$^{1/2}$, 3.0 m Ceiling Height, Radial Distance = 2.2 m

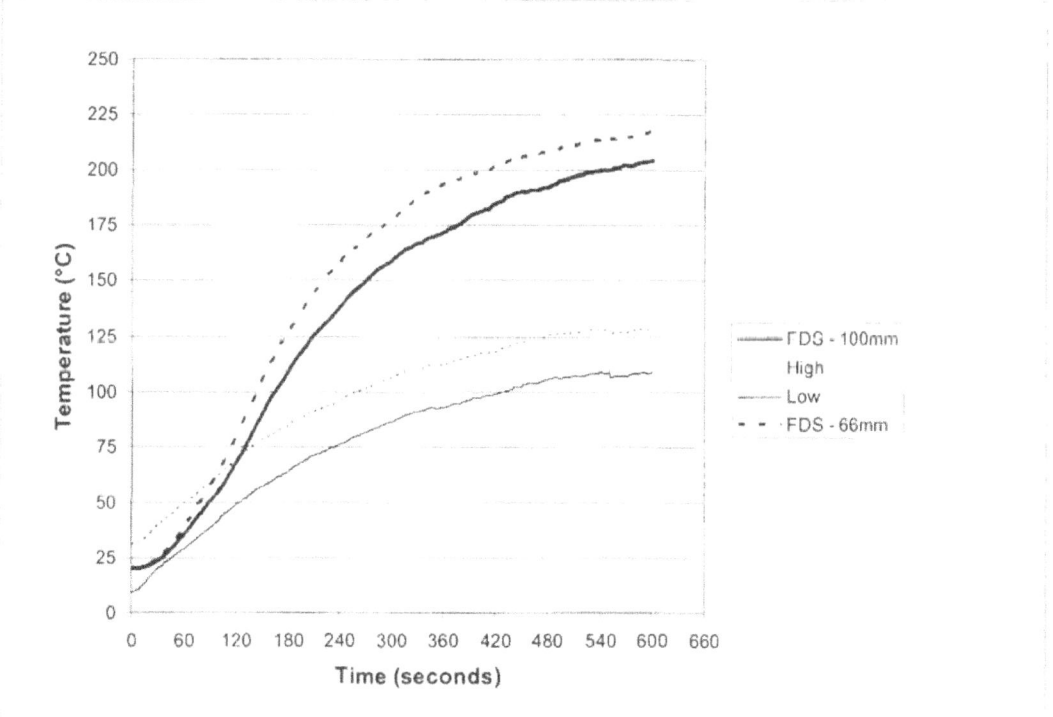

Figure B.8 – Comparison of Predicted and Measured Heat Detector Temperatures, RTI = 287 m$^{1/2}$-s$^{1/2}$, 3.0 m Ceiling Height, Radial Distance = 2.2 m

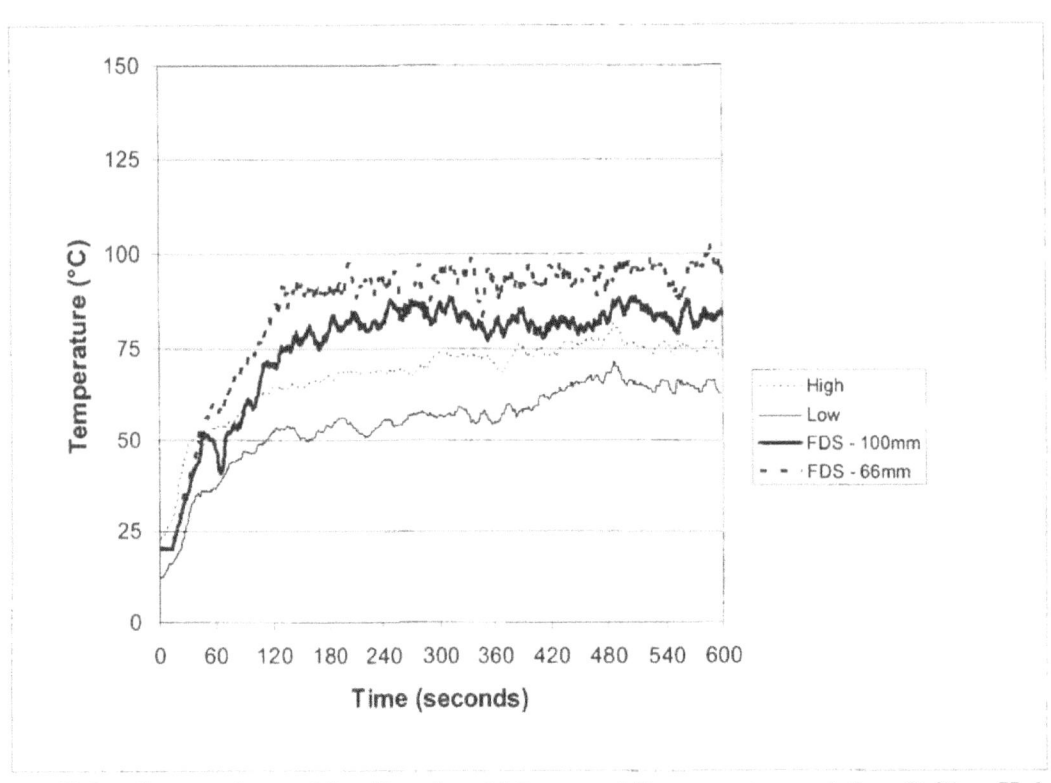

Figure B.9 – Comparison of Predicted and Measured Temperatures, 3.0 m Ceiling Height, Radial Distance = 6.5 m

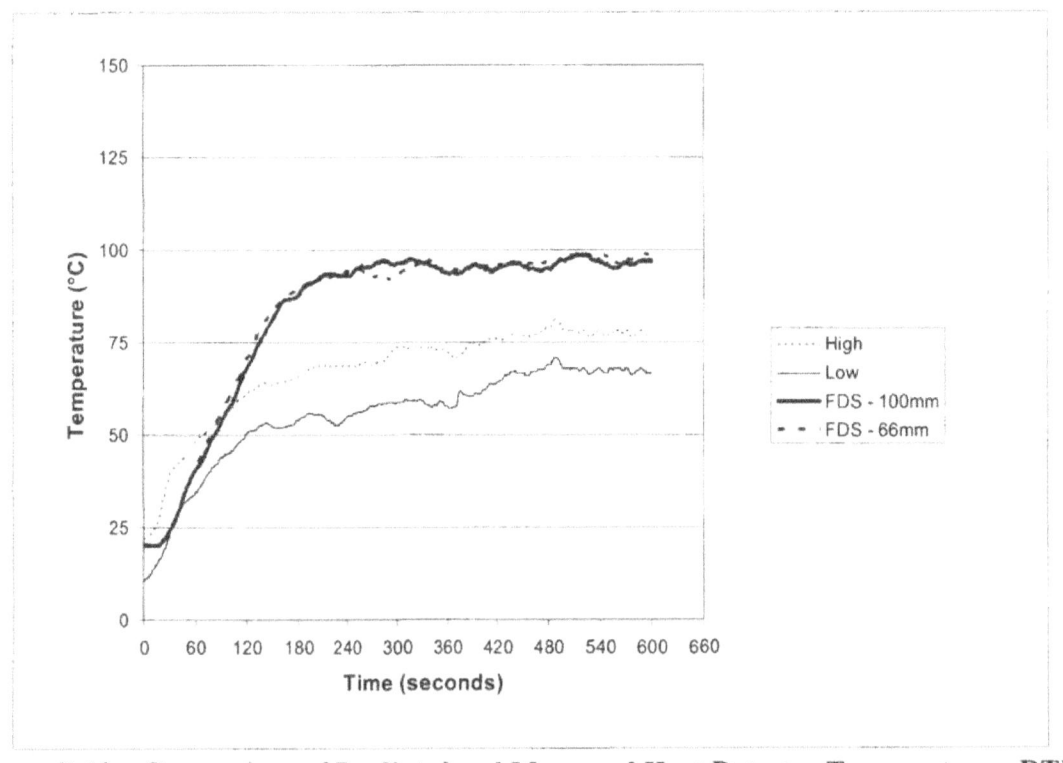

Figure B.10 – Comparison of Predicted and Measured Heat Detector Temperatures, RTI = 32 m$^{1/2}$-s$^{1/2}$, 3.0 m Ceiling Height, Radial Distance = 6.5 m

58

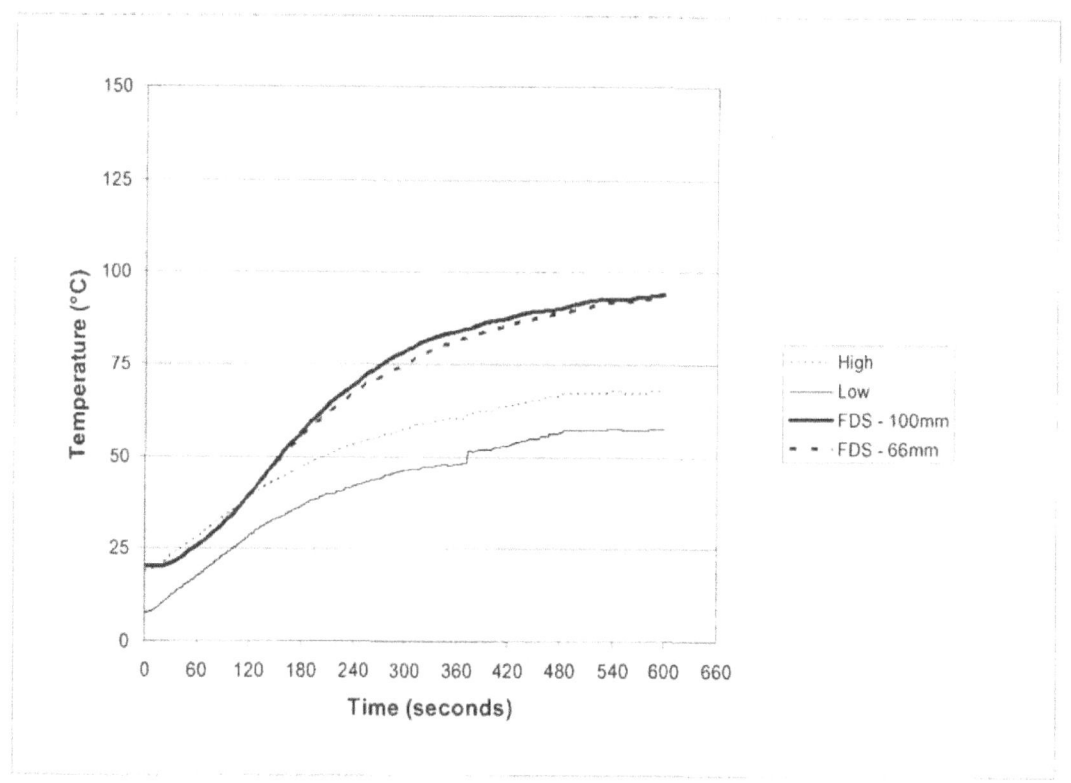

Figure B.11 – Comparison of Predicted and Measured Heat Detector Temperatures, RTI = 164 m$^{1/2}$-s$^{1/2}$, 3.0 m Ceiling Height, Radial Distance = 6.5 m

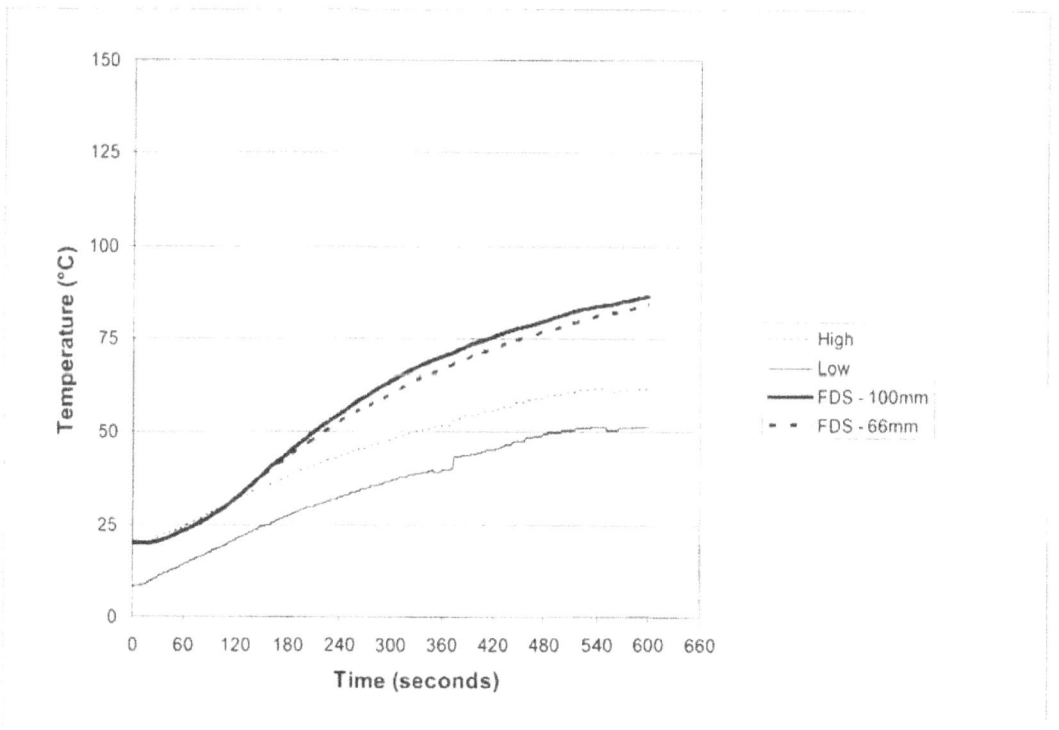

Figure B.12 – Comparison of Predicted and Measured Heat Detector Temperatures, RTI = 287 m$^{1/2}$-s$^{1/2}$, 3.0 m Ceiling Height, Radial Distance = 6.5 m

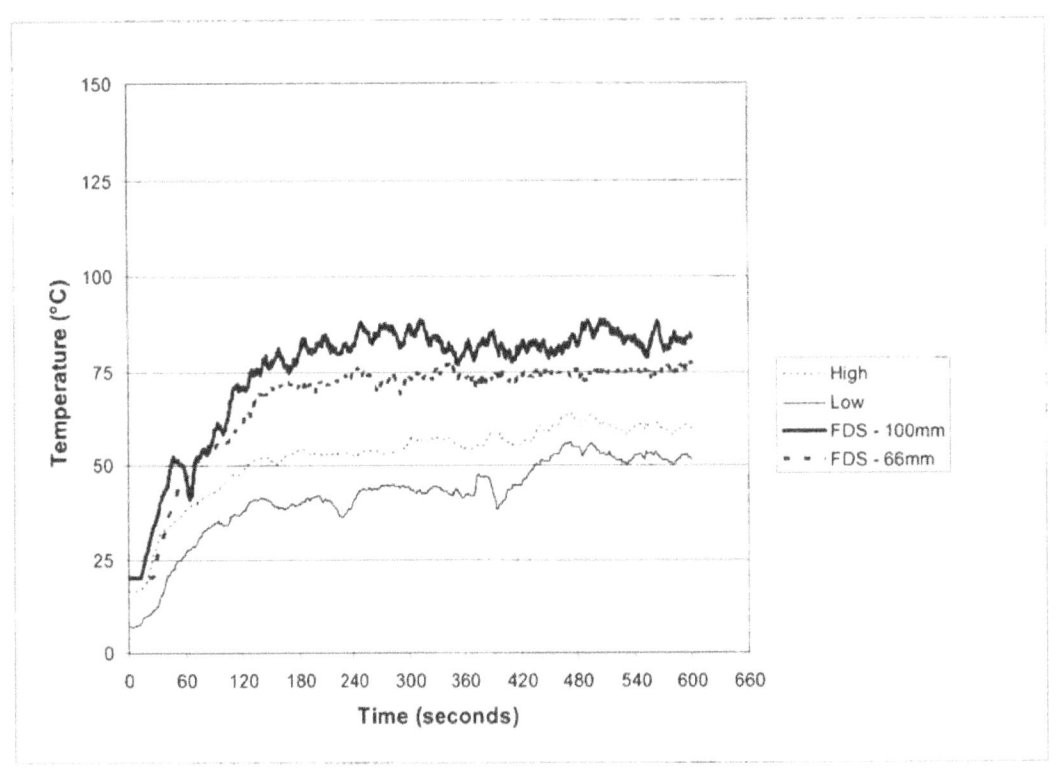

Figure B.13 – Comparison of Predicted and Measured Temperatures, 3.0 m Ceiling Height, Radial Distance = 10.8 m

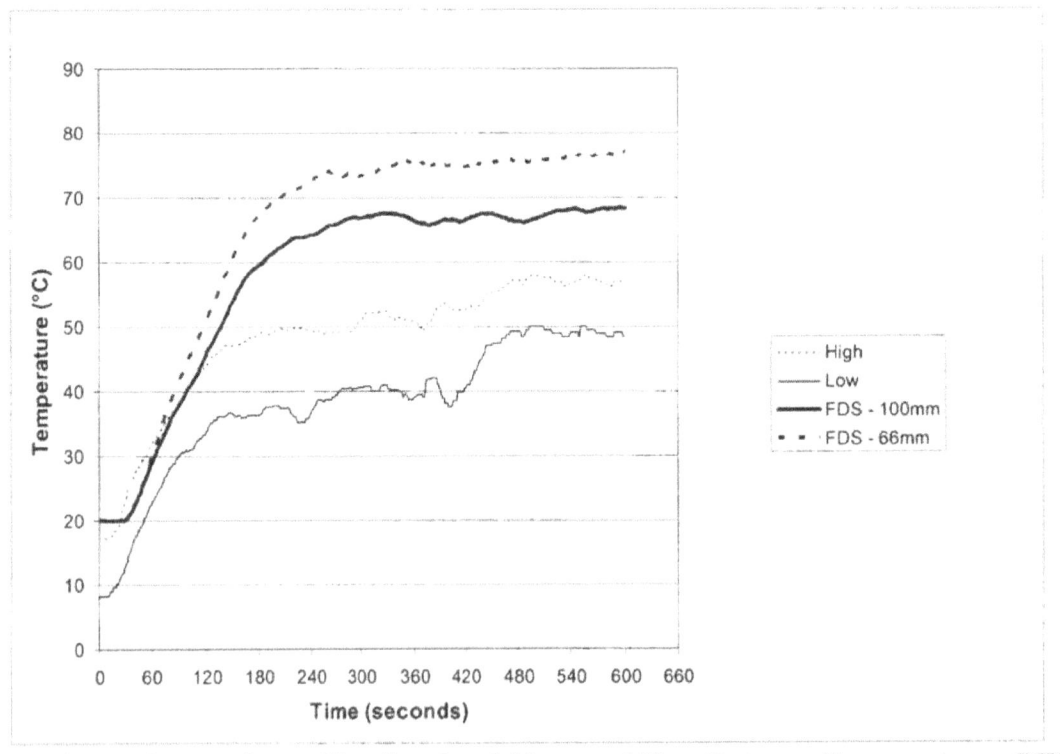

Figure B.14 – Comparison of Predicted and Measured Heat Detector Temperatures, RTI = 32 $m^{1/2}$-$s^{1/2}$, 3.0 m Ceiling Height, Radial Distance = 10.8 m

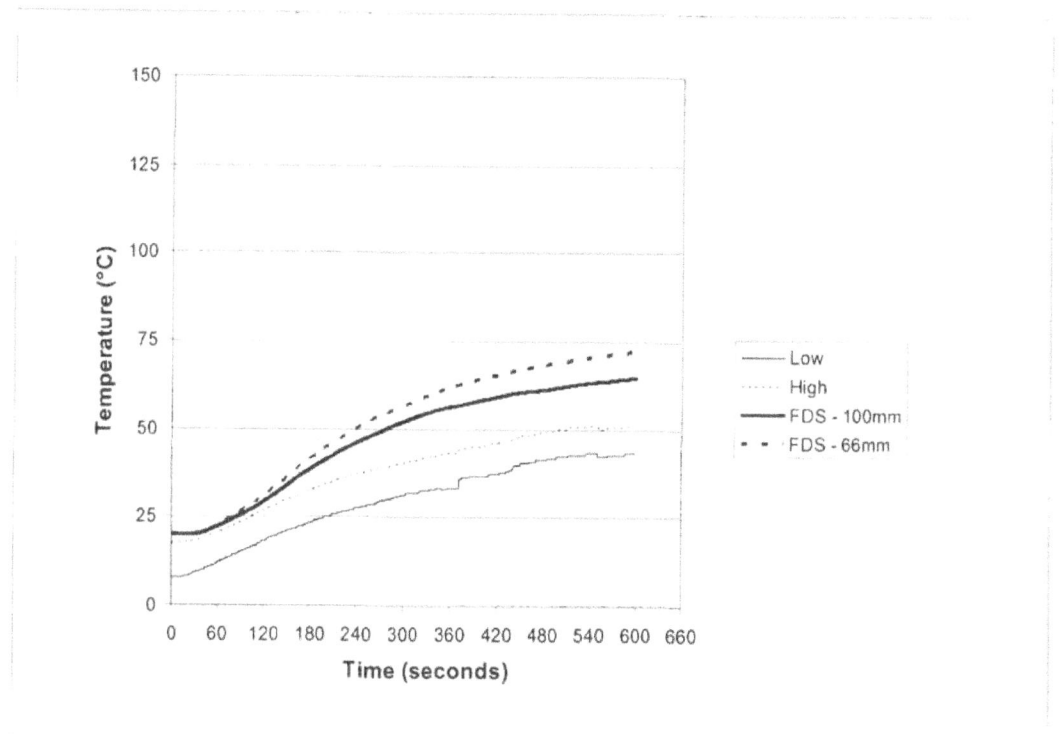

Figure B.15 – Comparison of Predicted and Measured Heat Detector Temperatures, RTI = 164 $m^{1/2}$-$s^{1/2}$, 3.0 m Ceiling Height, Radial Distance = 10.8 m

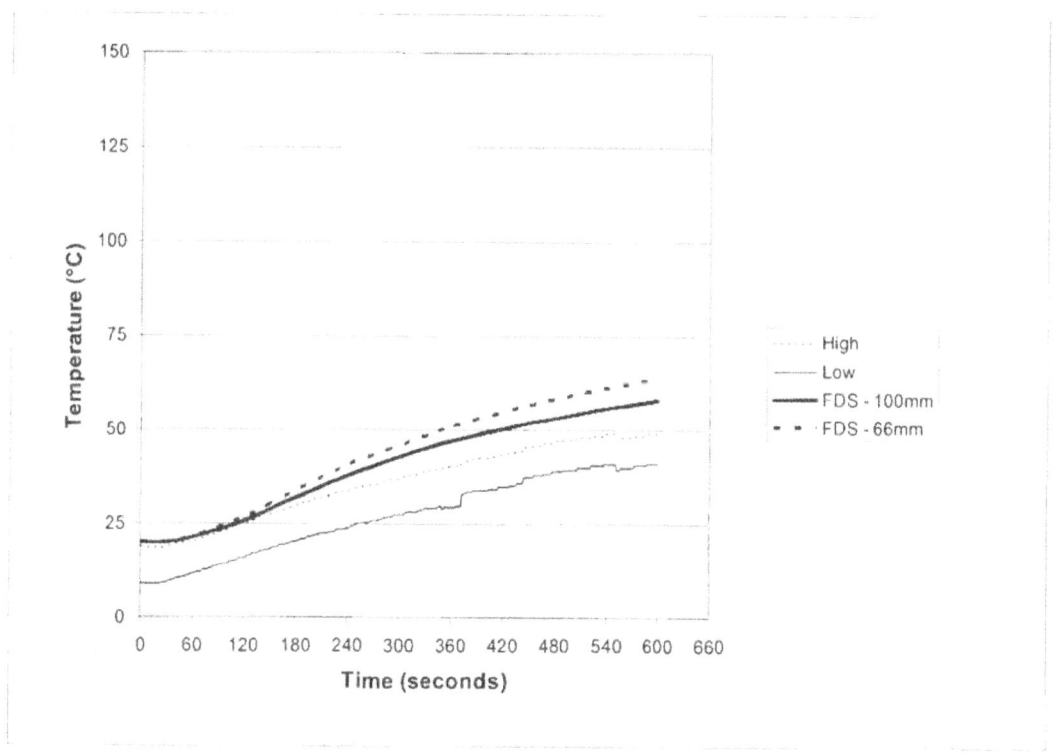

Figure B.16 – Comparison of Predicted and Measured Heat Detector Temperatures, RTI = 287 $m^{1/2}$-$s^{1/2}$, 3.0 m Ceiling Height, Radial Distance = 10.8 m

61

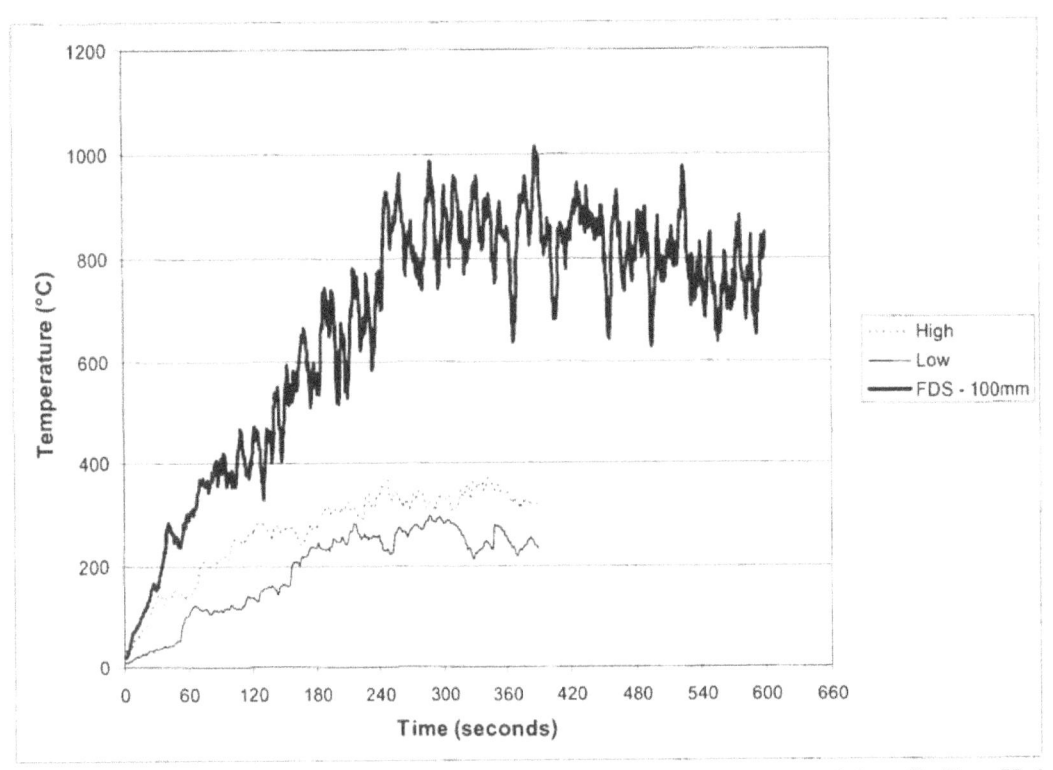

Figure B.17 – Comparison of Predicted and Measured Temperatures, 4.6 m Ceiling Height, Plume Centerline

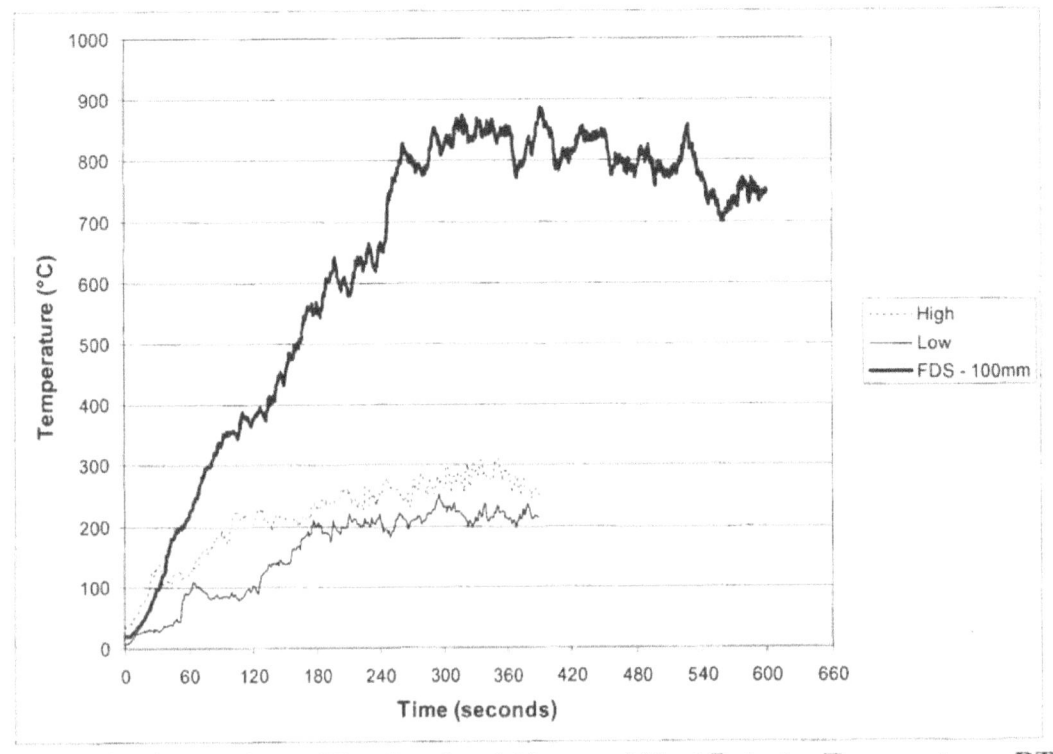

Figure B.18 – Comparison of Predicted and Measured Heat Detector Temperatures, RTI = 32 m$^{1/2}$-s$^{1/2}$, 4.6 m Ceiling Height, Plume Centerline

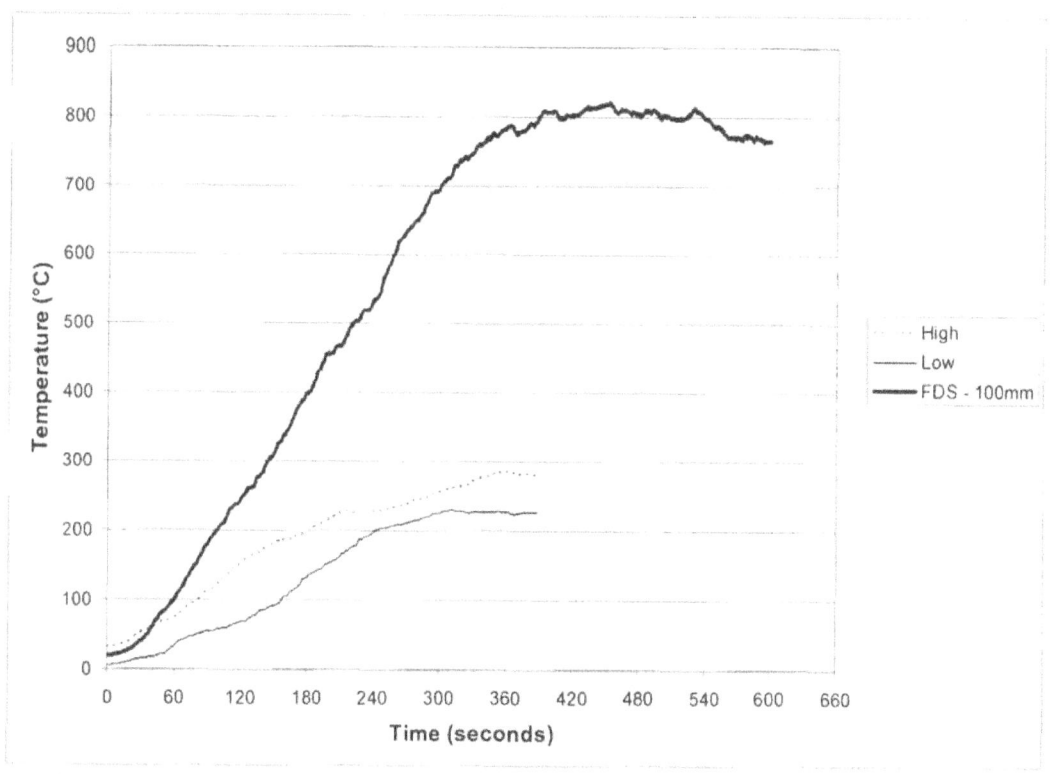

Figure B.19 – Comparison of Predicted and Measured Heat Detector Temperatures, RTI = 164 m$^{1/2}$-s$^{1/2}$, 4.6 m Ceiling Height, Plume Centerline

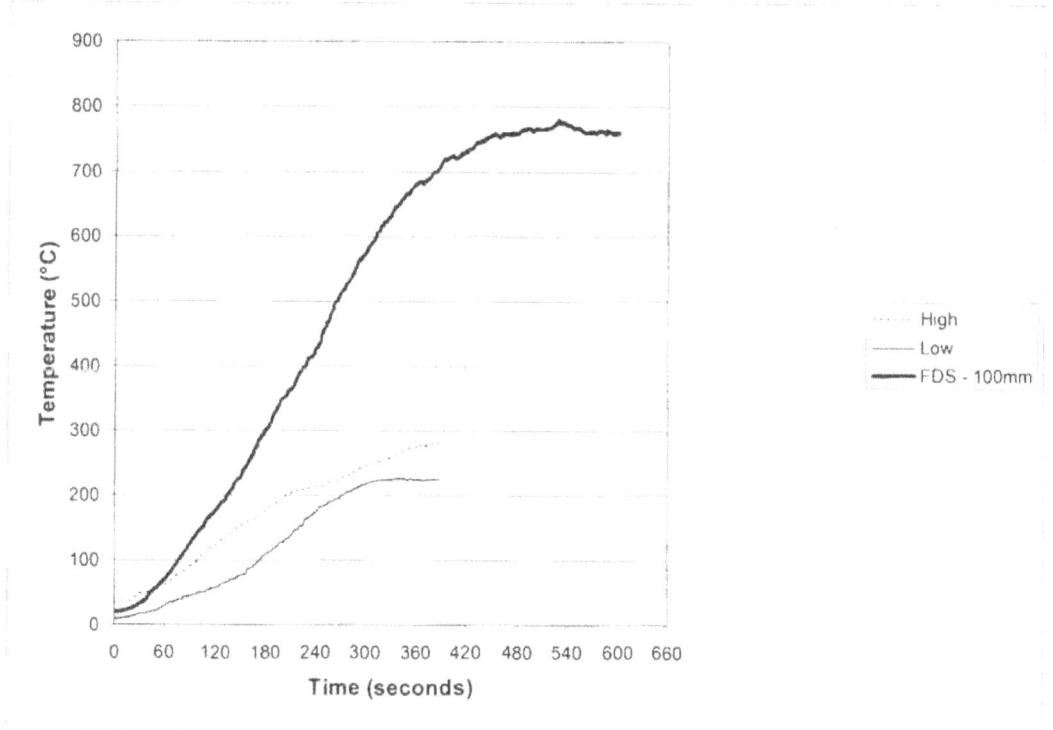

Figure B.20 – Comparison of Predicted and Measured Heat Detector Temperatures, RTI = 287 m$^{1/2}$-s$^{1/2}$, 4.6 m Ceiling Height, Plume Centerline

63

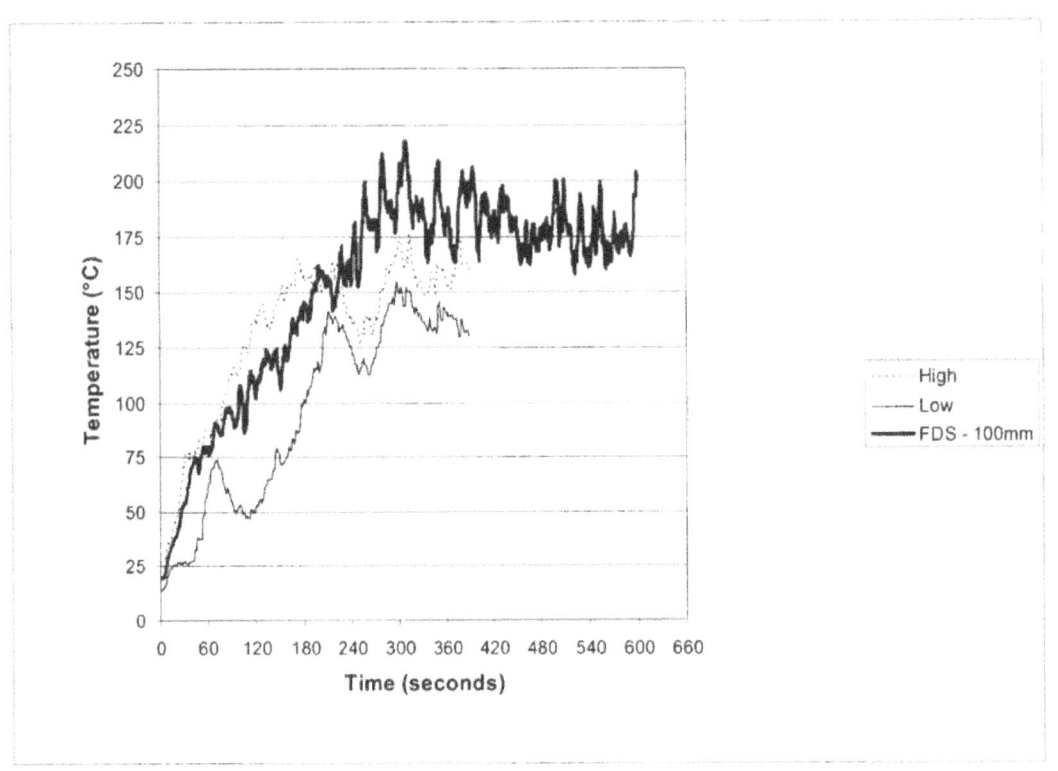

Figure B.21 – Comparison of Predicted and Measured Temperatures, 4.6 m Ceiling Height, Radial Distance = 2.2 m

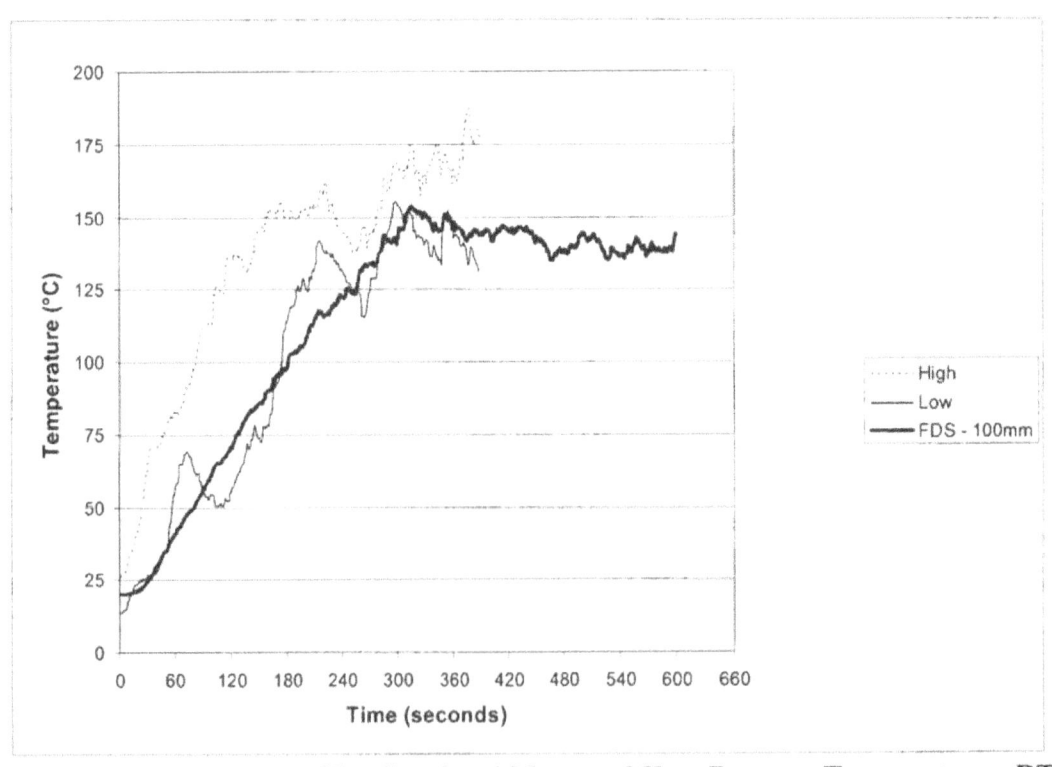

Figure B.22 – Comparison of Predicted and Measured Heat Detector Temperatures, RT1 = 32 m$^{1/2}$-s$^{1/2}$, 4.6 m Ceiling Height, Radial Distance = 2.2 m

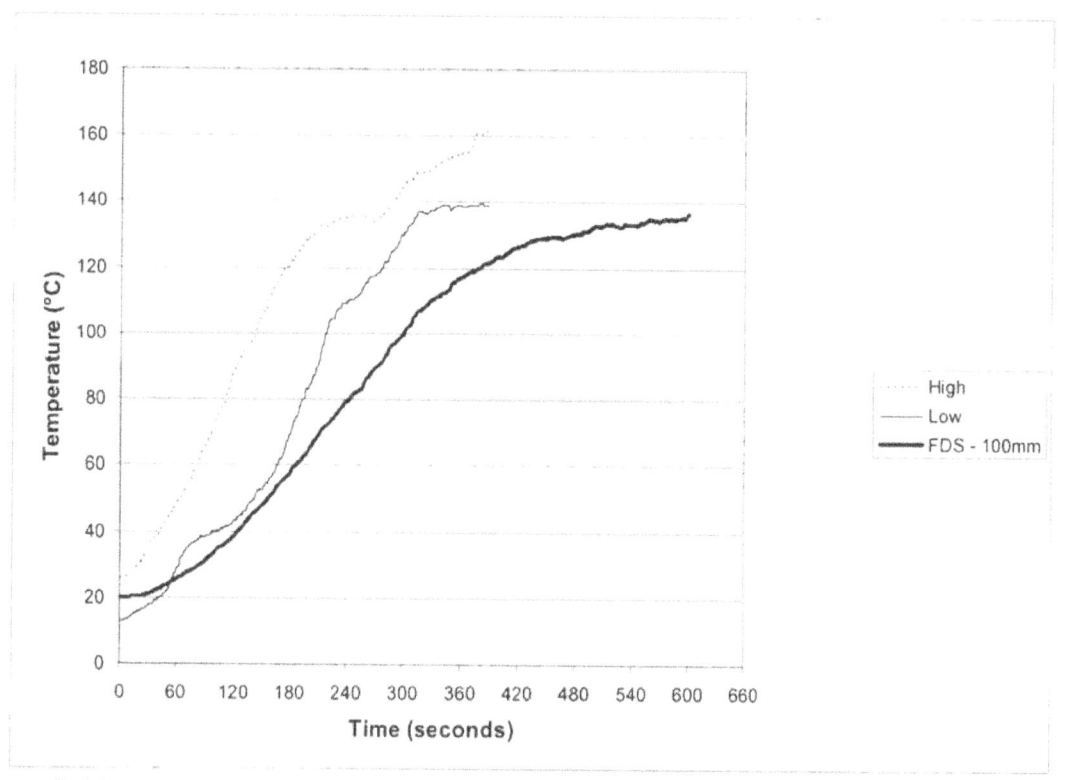

Figure B.23 – Comparison of Predicted and Measured Heat Detector Temperatures, RTI = 164 m$^{1/2}$-s$^{1/2}$, 4.6 m Ceiling Height, Radial Distance = 2.2 m

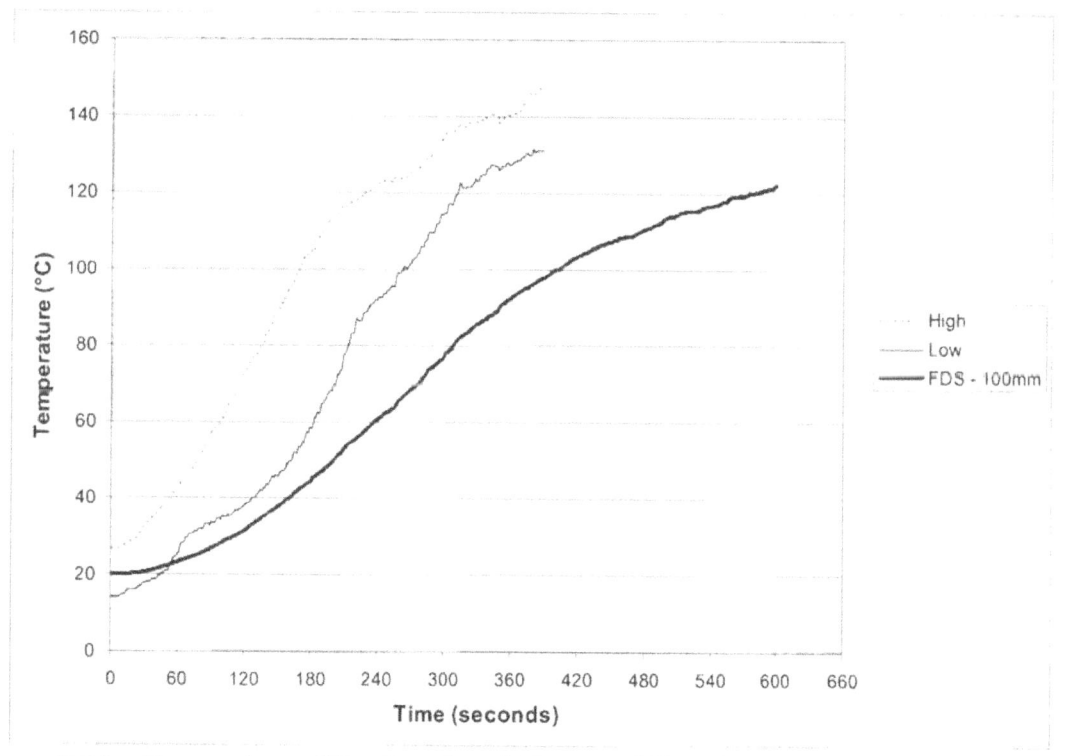

Figure B.24 – Comparison of Predicted and Measured Heat Detector Temperatures, RTI = 287 m$^{1/2}$-s$^{1/2}$, 4.6 m Ceiling Height, Radial Distance = 2.2 m

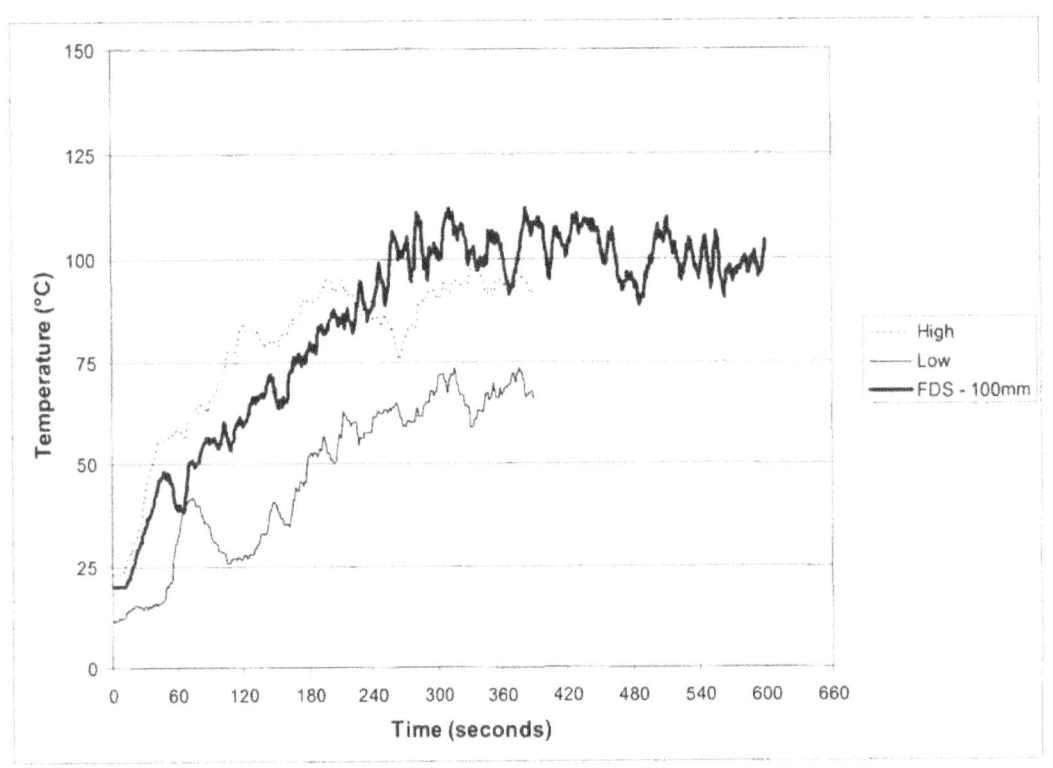

Figure B.25 – Comparison of Predicted and Measured Temperatures, 4.6 m Ceiling Height, Radial Distance = 6.5 m

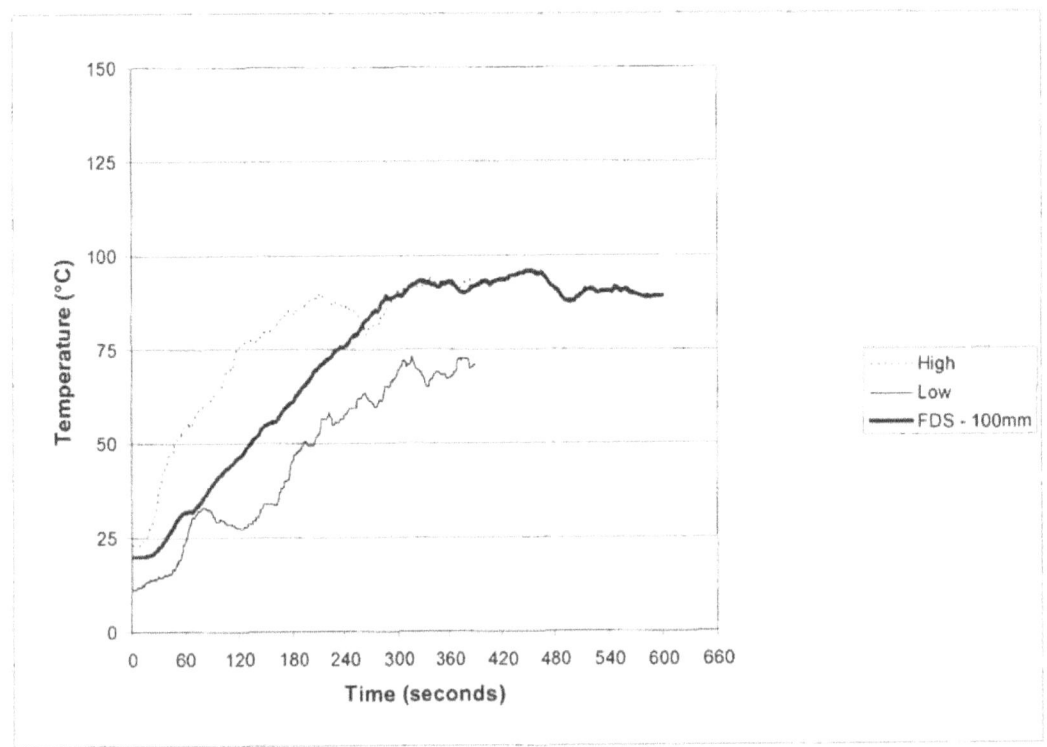

Figure B.26 – Comparison of Predicted and Measured Heat Detector Temperatures, RTI = 32 m$^{1/2}$-s$^{1/2}$, 4.6 m Ceiling Height, Radial Distance = 6.5 m

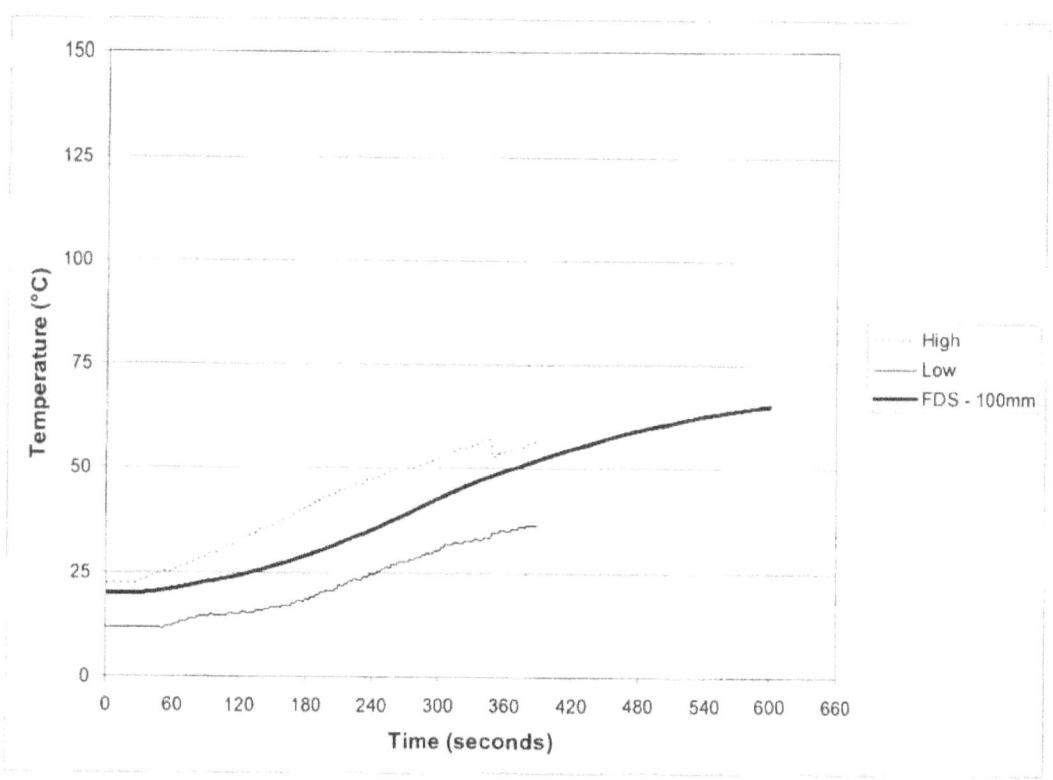

Figure B.27 – Comparison of Predicted and Measured Heat Detector Temperatures, RTI = 164 m$^{1/2}$-s$^{1/2}$, 4.6 m Ceiling Height, Radial Distance = 6.5 m

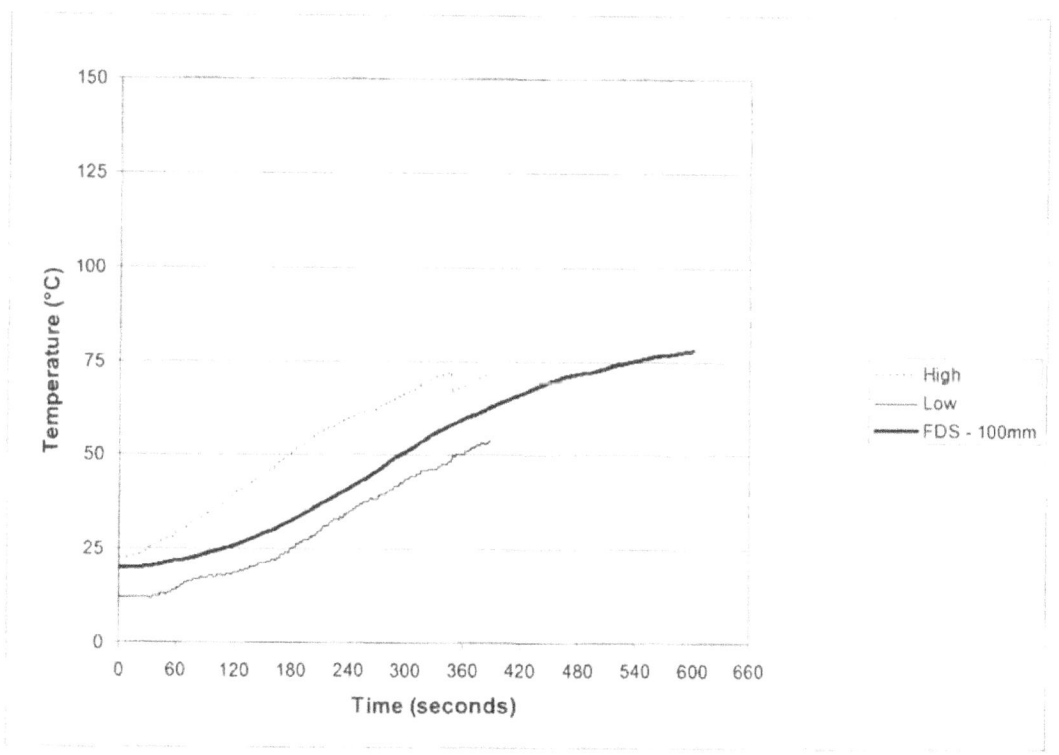

Figure B.28 – Comparison of Predicted and Measured Heat Detector Temperatures, RTI = 287 m$^{1/2}$-s$^{1/2}$, 4.6 m Ceiling Height, Radial Distance = 6.5 m

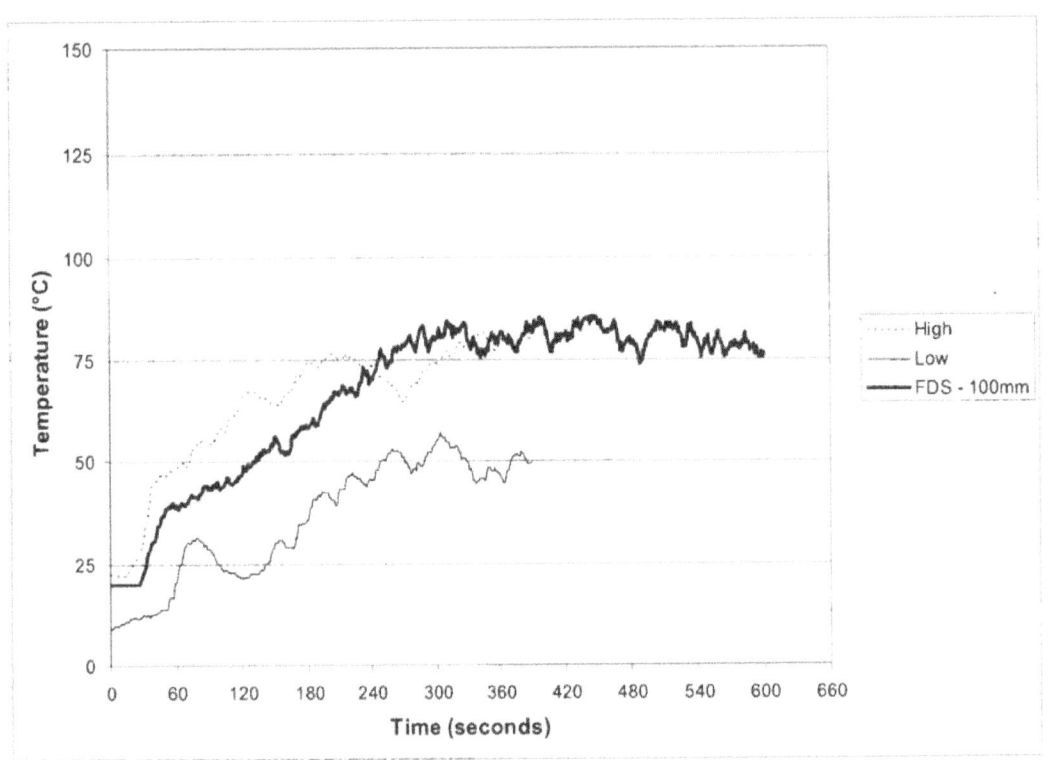

Figure B.29 – Comparison of Predicted and Measured Temperatures, 4.6 m Ceiling Height, Radial Distance = 10.8 m

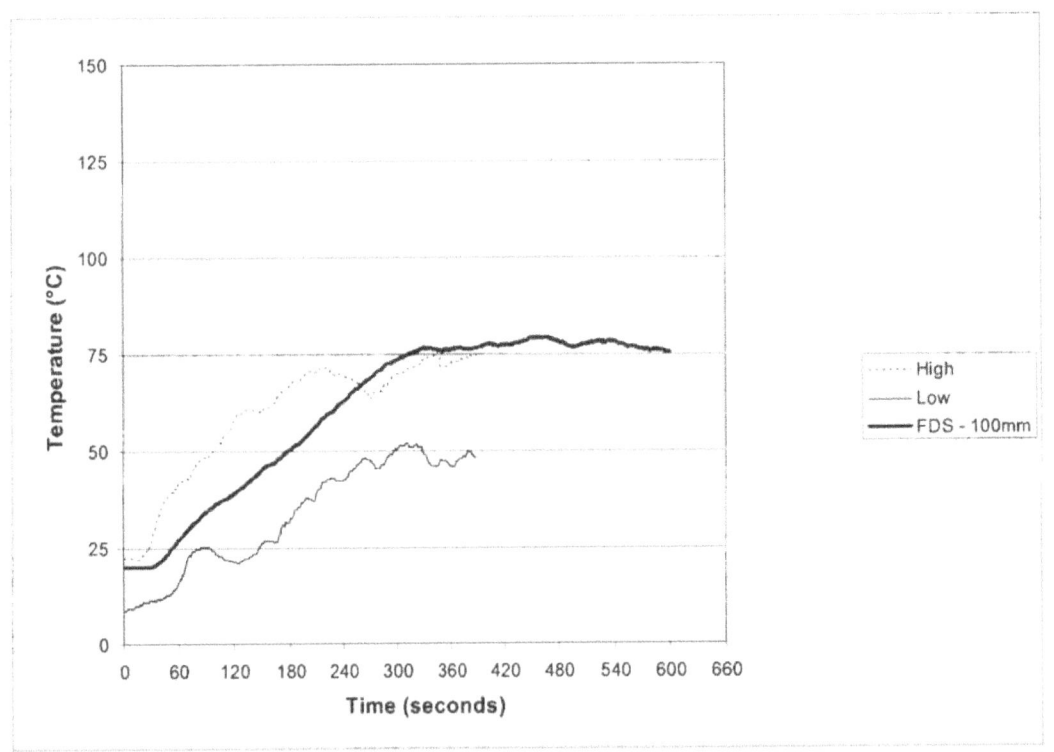

Figure B.30 – Comparison of Predicted and Measured Heat Detector Temperatures, RTI = 32 m$^{1/2}$-s$^{1/2}$, 4.6 m Ceiling Height, Radial Distance = 10.8 m

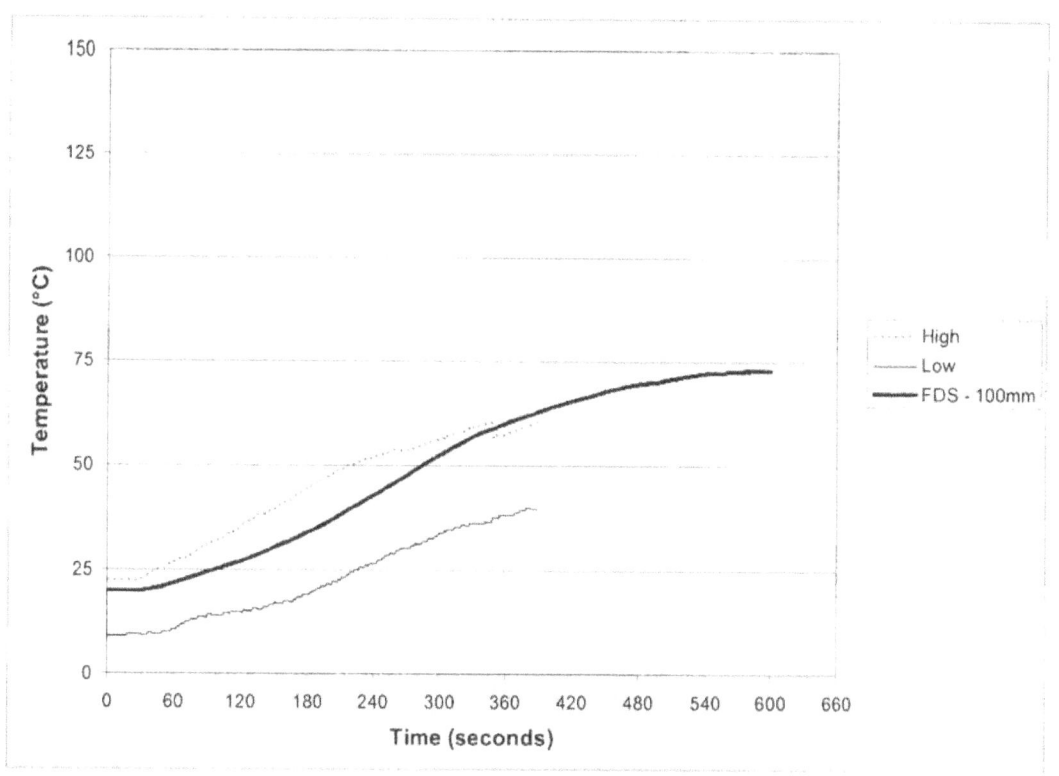

Figure B.31 – Comparison of Predicted and Measured Heat Detector Temperatures, RTI = 164 $m^{1/2}$-$s^{1/2}$, 4.6 m Ceiling Height, Radial Distance = 10.8 m

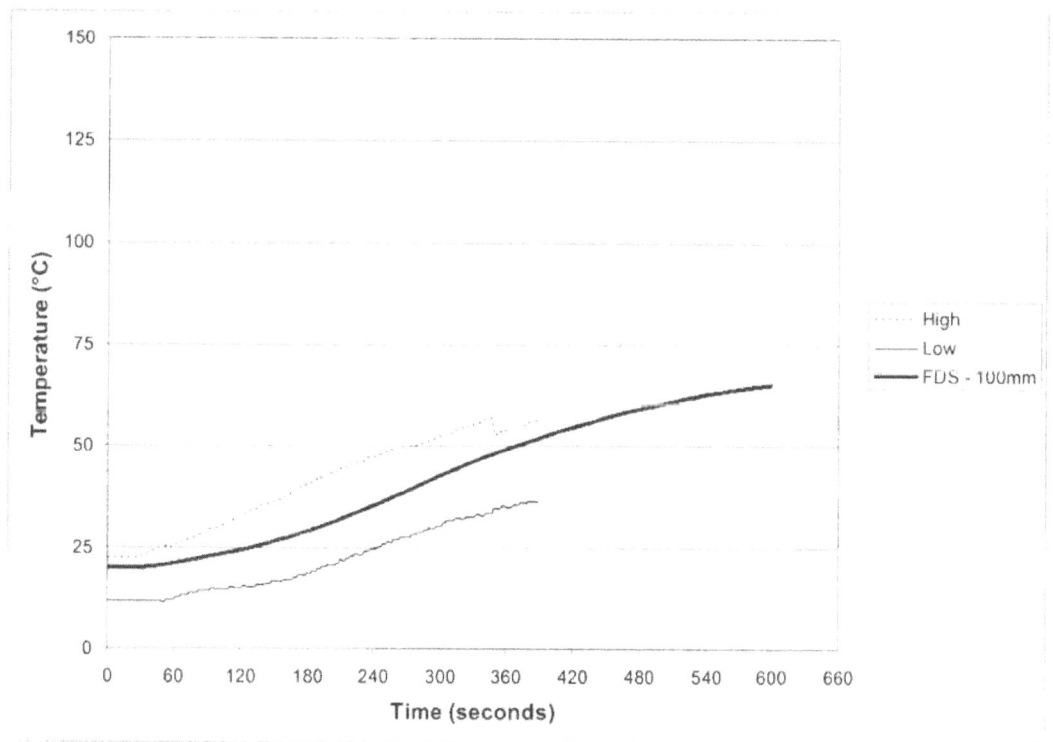

Figure B.32 – Comparison of Predicted and Measured Heat Detector Temperatures, RTI = 287 $m^{1/2}$-$s^{1/2}$, 4.6 m Ceiling Height, Radial Distance = 10.8 m

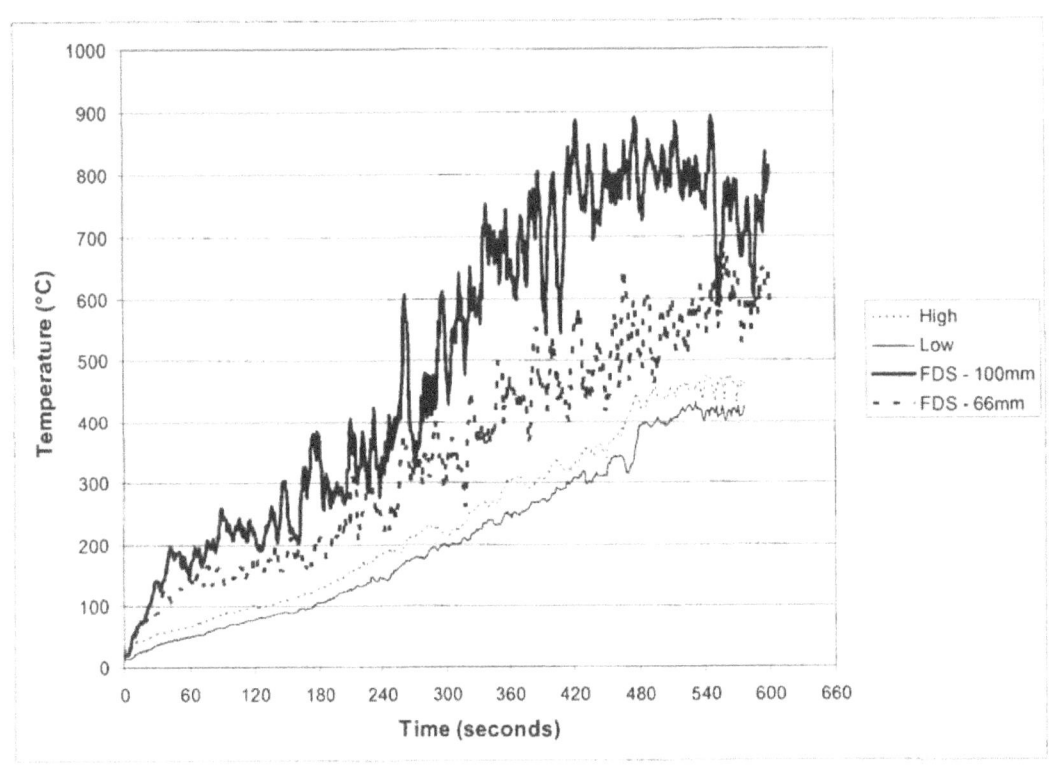

Figure B.33 – Comparison of Predicted and Measured Temperatures, 6.1 m Ceiling Height, Plume Centerline

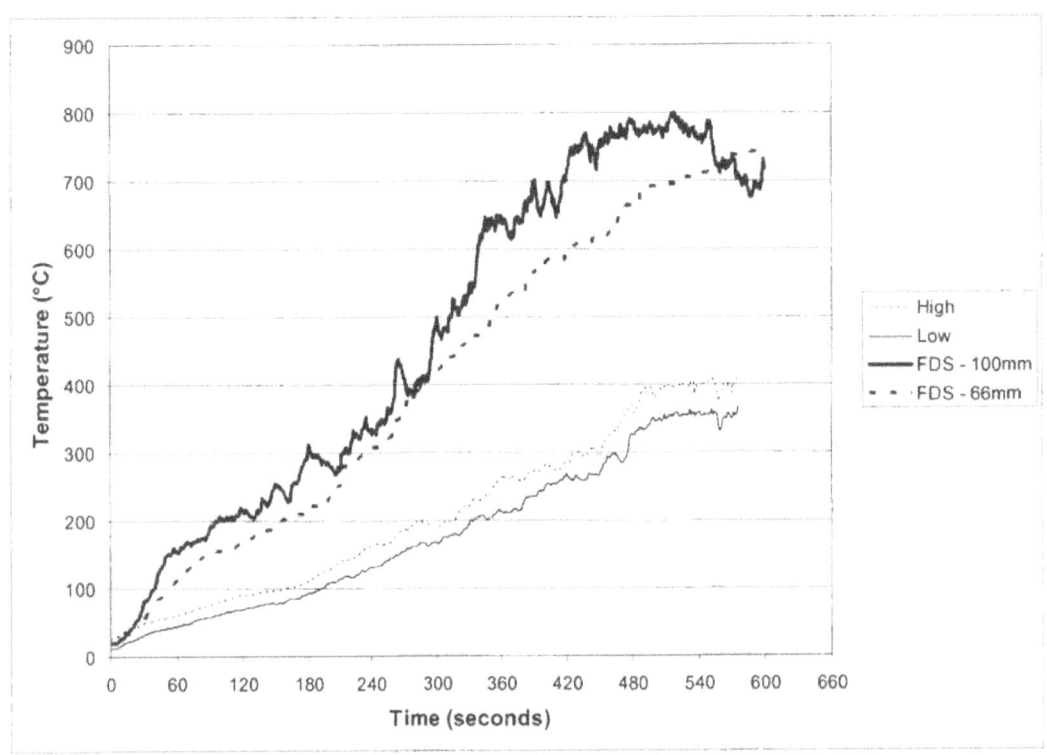

Figure B.34 – Comparison of Predicted and Measured Heat Detector Temperatures, RTI = 32 m$^{1/2}$-s$^{1/2}$, 6.1 m Ceiling Height, Plume Centerline

70

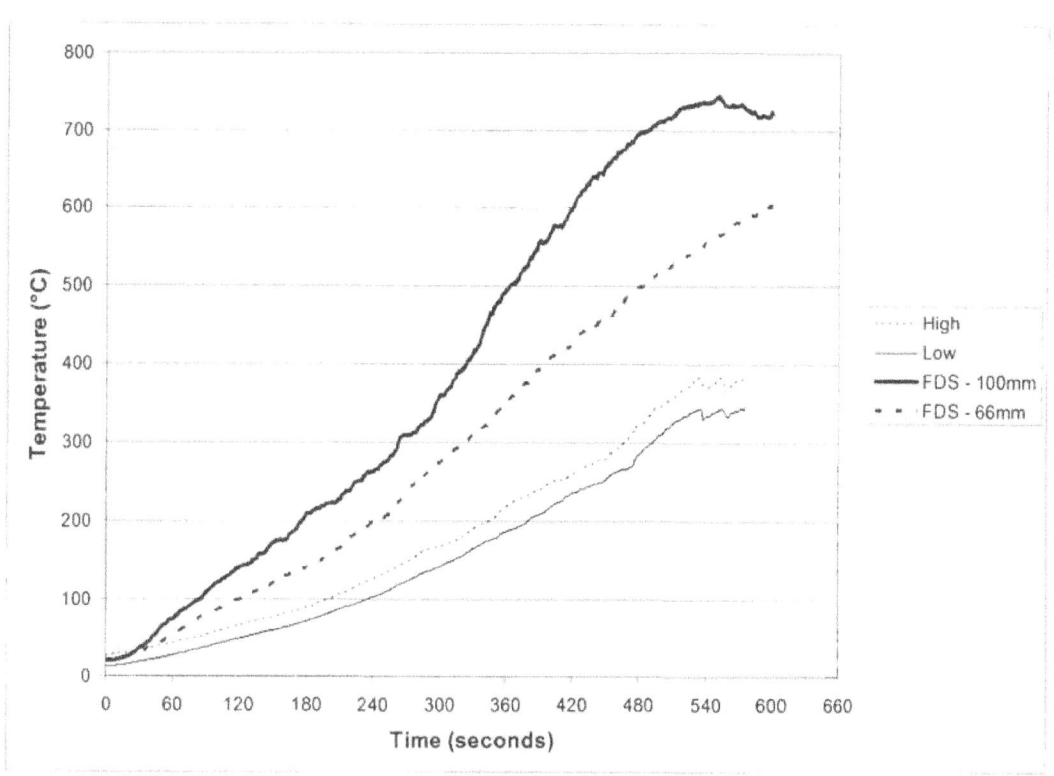

Figure B.35 – Comparison of Predicted and Measured Heat Detector Temperatures, RTI = 164 m$^{1/2}$-s$^{1/2}$, 6.1 m Ceiling Height, Plume Centerline

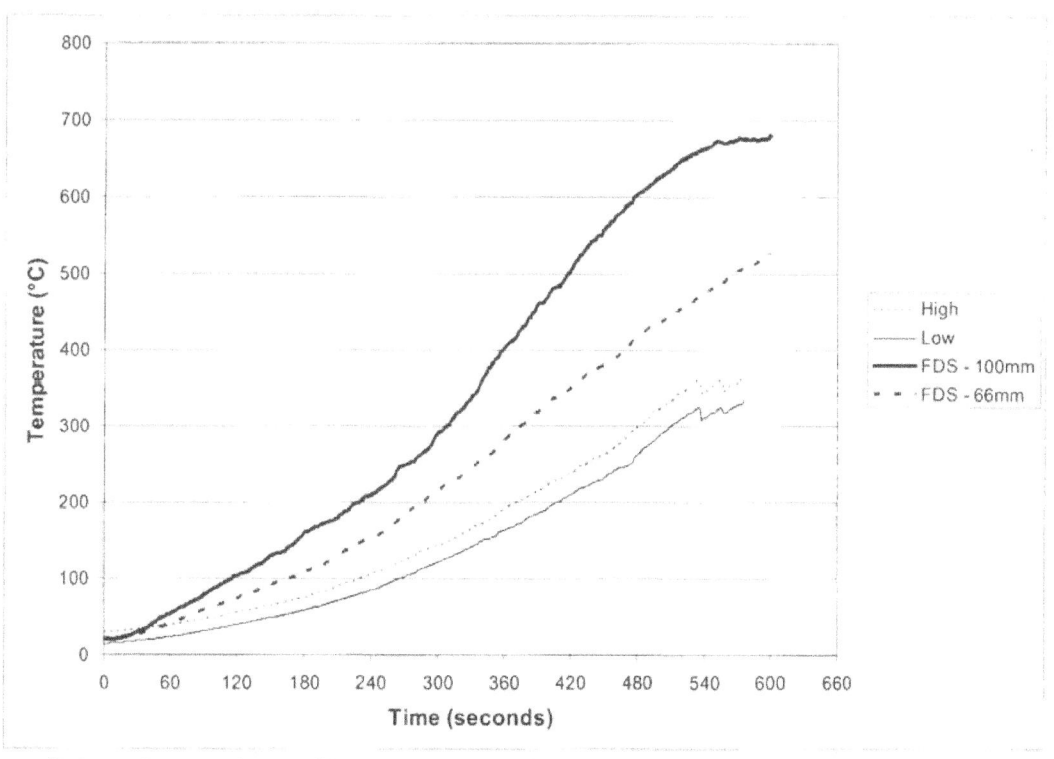

Figure B.36 – Comparison of Predicted and Measured Heat Detector Temperatures, RTI = 287 m$^{1/2}$-s$^{1/2}$, 6.1 m Ceiling Height, Plume Centerline

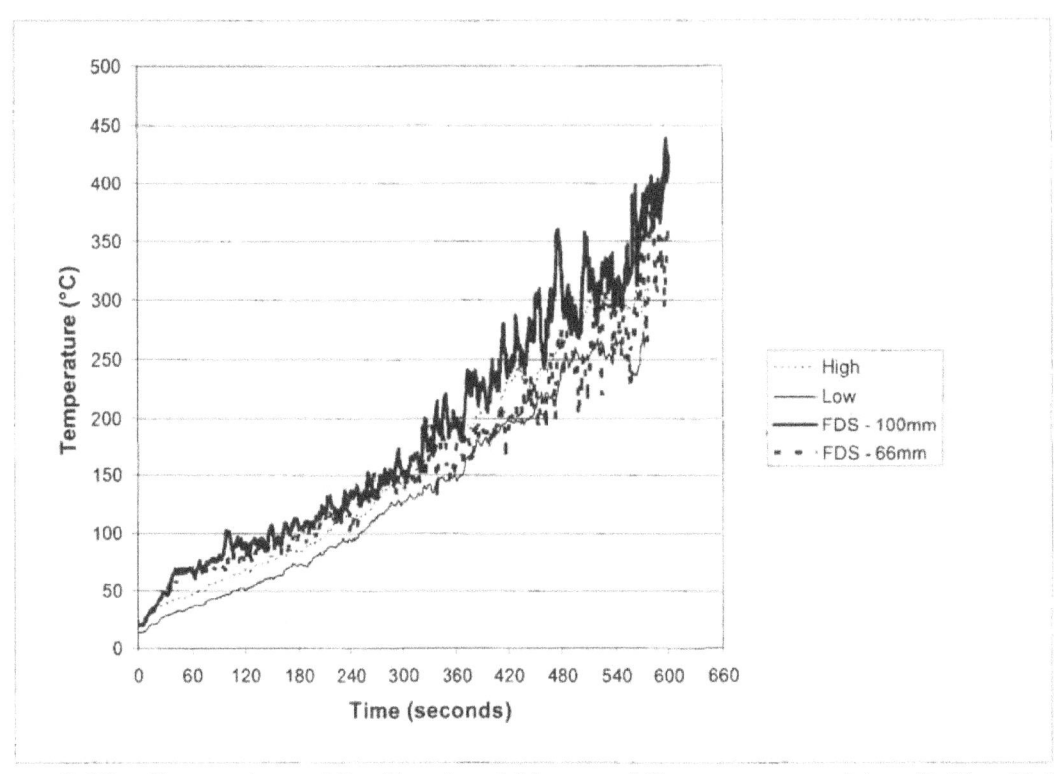

Figure B.37 – Comparison of Predicted and Measured Temperatures, 6.1 m Ceiling Height, Radial Distance = 2.2 m

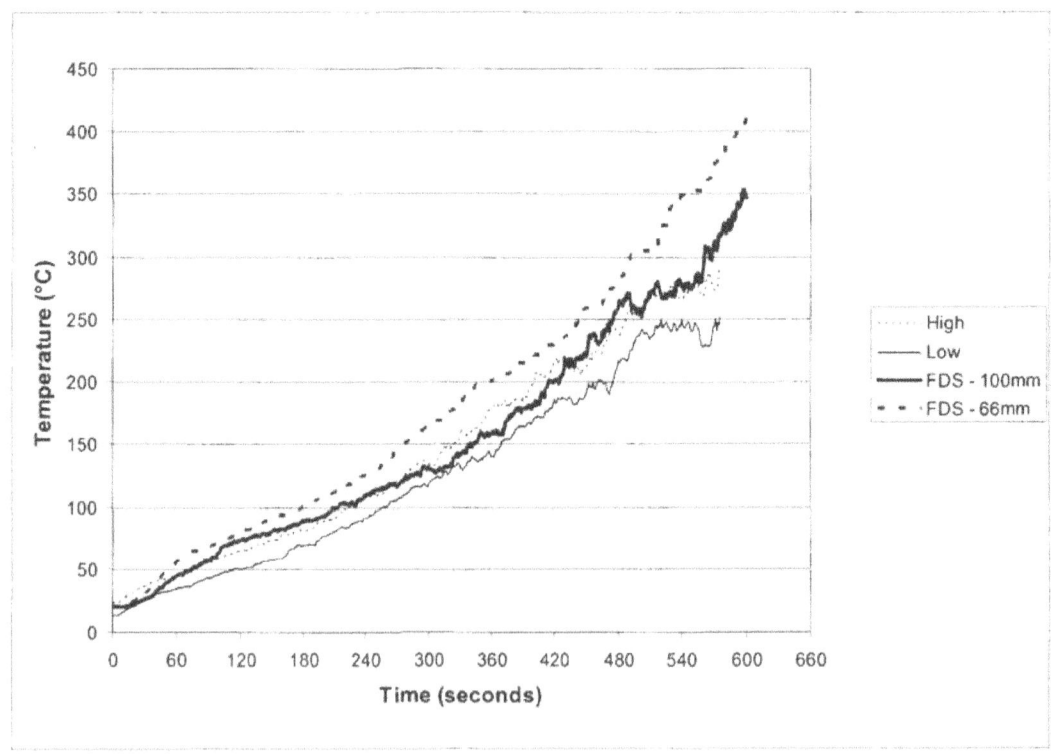

Figure B.38 – Comparison of Predicted and Measured Heat Detector Temperatures, RTI = 32 m$^{1/2}$-s$^{1/2}$, 6.1 m Ceiling Height, Radial Distance = 2.2 m

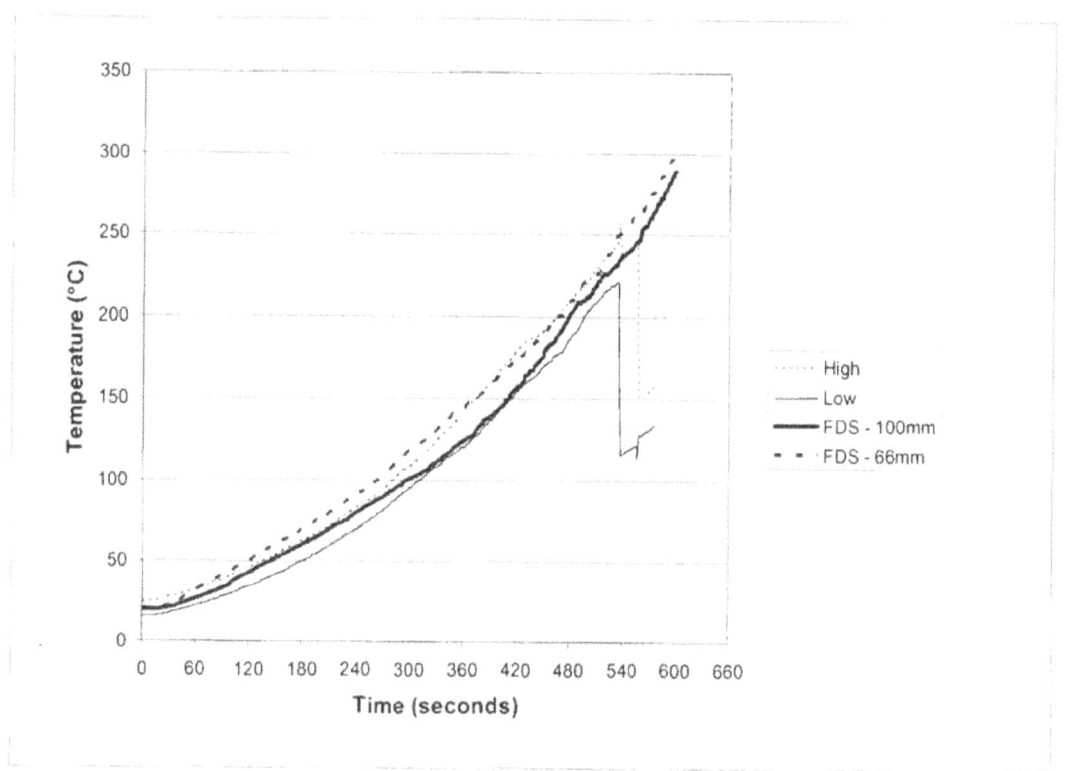

Figure B.39 – Comparison of Predicted and Measured Heat Detector Temperatures, RTI = 164 $m^{1/2}$-$s^{1/2}$, 6.1 m Ceiling Height, Radial Distance = 2.2 m

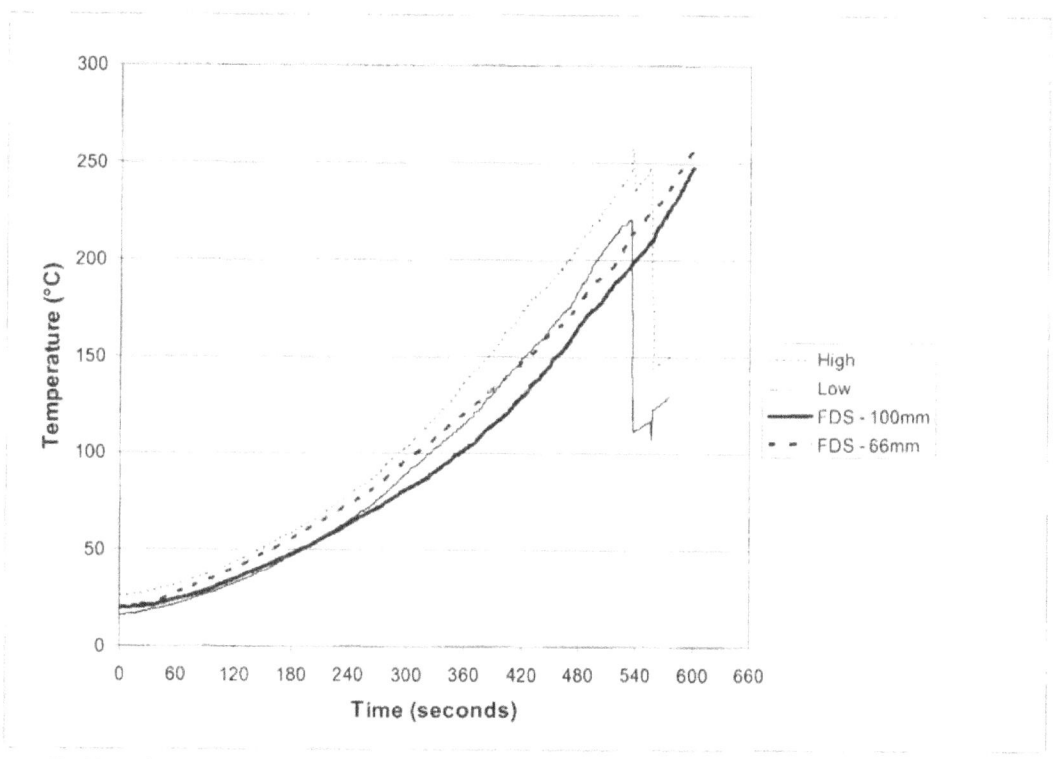

Figure B.40 – Comparison of Predicted and Measured Heat Detector Temperatures, RTI = 287 $m^{1/2}$-$s^{1/2}$, 6.1 m Ceiling Height, Radial Distance = 2.2 m

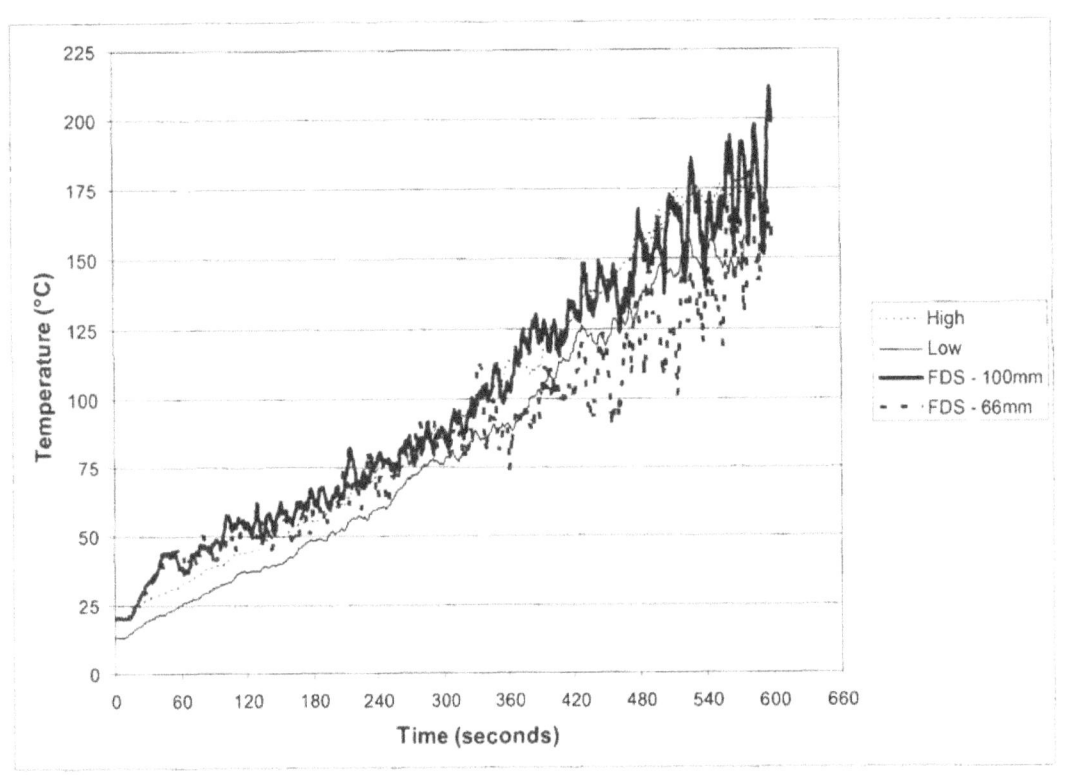

Figure B.41 – Comparison of Predicted and Measured Temperatures, 6.1 m Ceiling Height, Radial Distance = 6.5 m

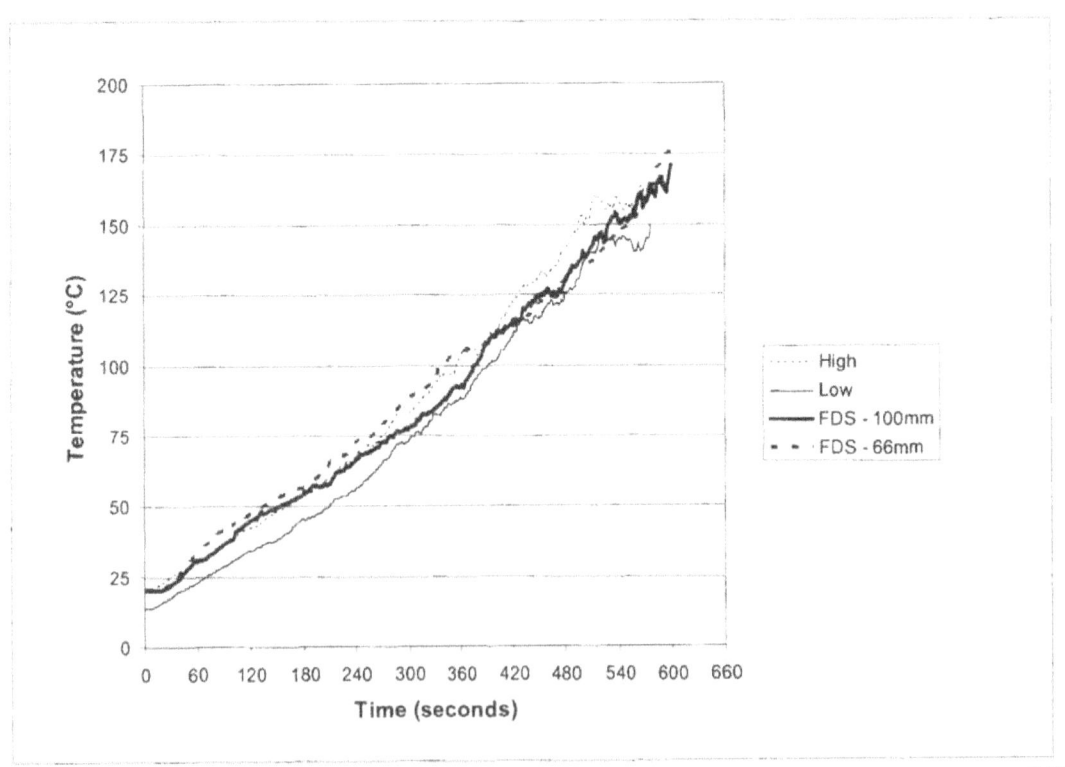

Figure B.42 – Comparison of Predicted and Measured Heat Detector Temperatures, RT1 = 32 m$^{1/2}$-s$^{1/2}$, 6.1 m Ceiling Height, Radial Distance = 6.5 m

74

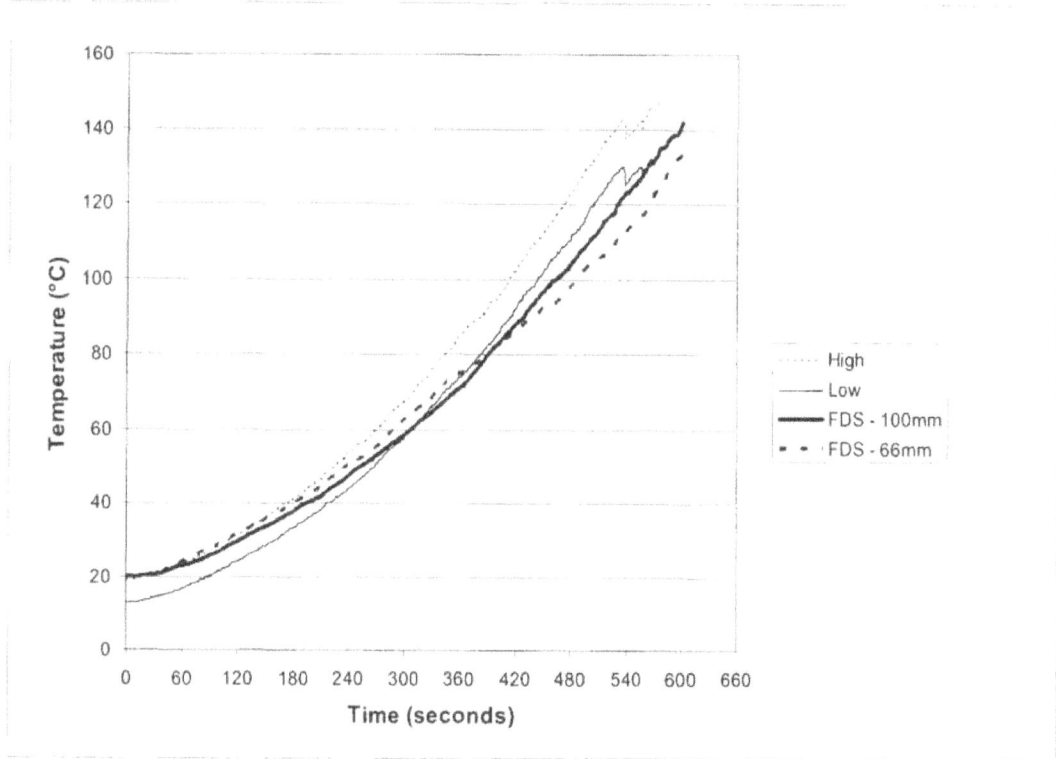

Figure B.43 – Comparison of Predicted and Measured Heat Detector Temperatures, RTI = 164 m$^{1/2}$-s$^{1/2}$, 6.1 m Ceiling Height, Radial Distance = 6.5 m

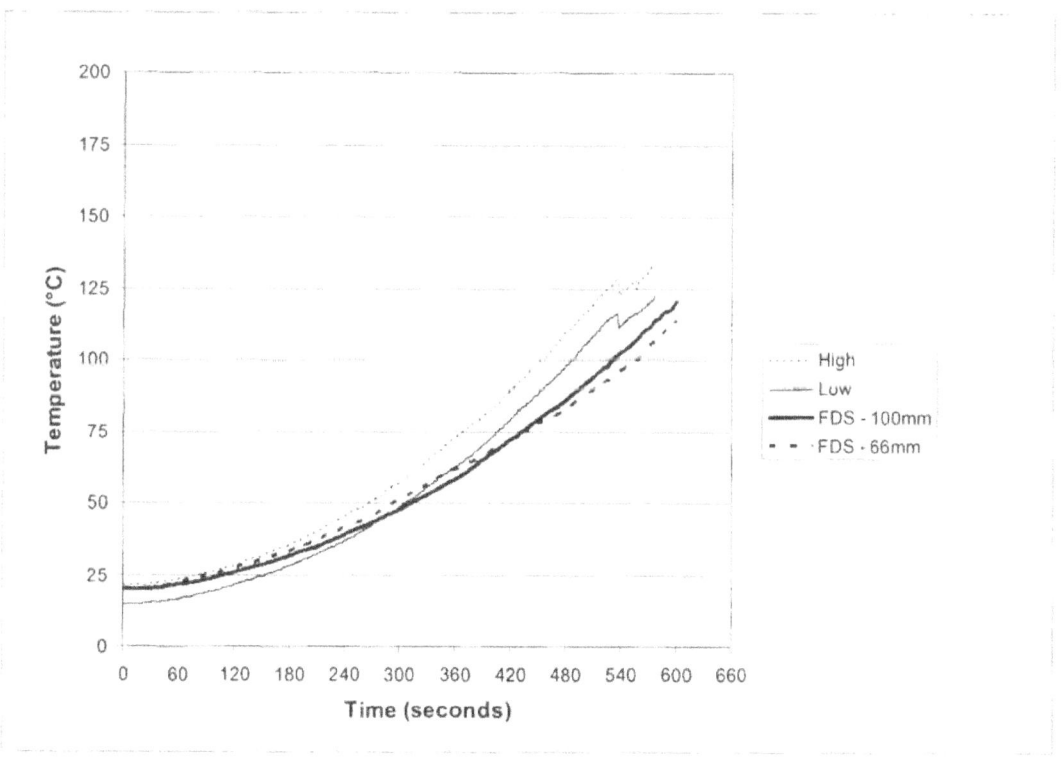

Figure B.44 – Comparison of Predicted and Measured Heat Detector Temperatures, RTI = 287 m$^{1/2}$-s$^{1/2}$, 6.1 m Ceiling Height, Radial Distance = 6.5 m

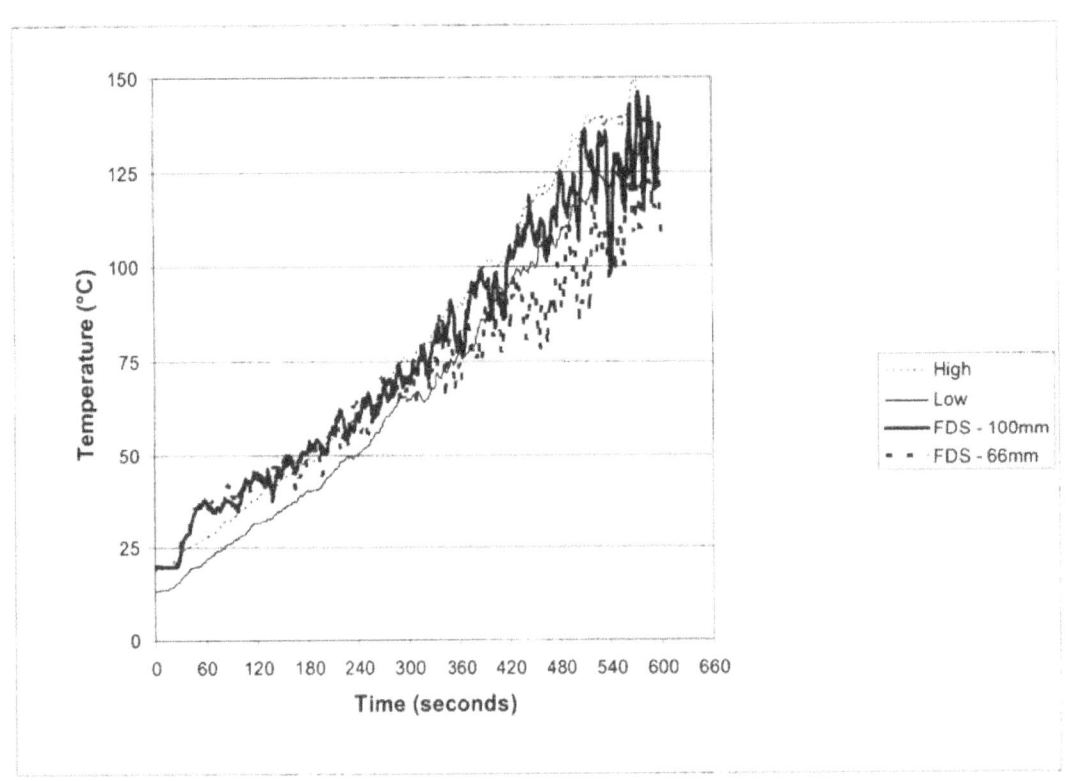

Figure B.45 – Comparison of Predicted and Measured Temperatures, 6.1 m Ceiling Height, Radial Distance = 10.8 m

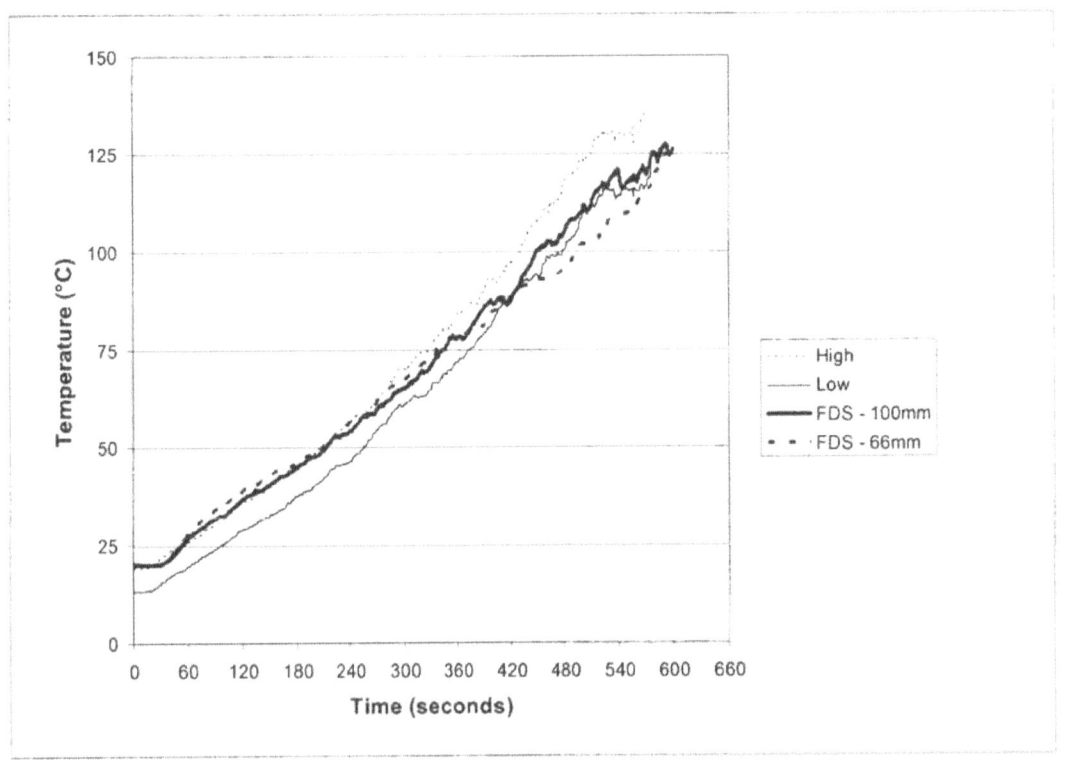

Figure B.46 – Comparison of Predicted and Measured Heat Detector Temperatures, RTl = 32 m$^{1/2}$-s$^{1/2}$, 6.1 m Ceiling Height, Radial Distance = 10.8 m

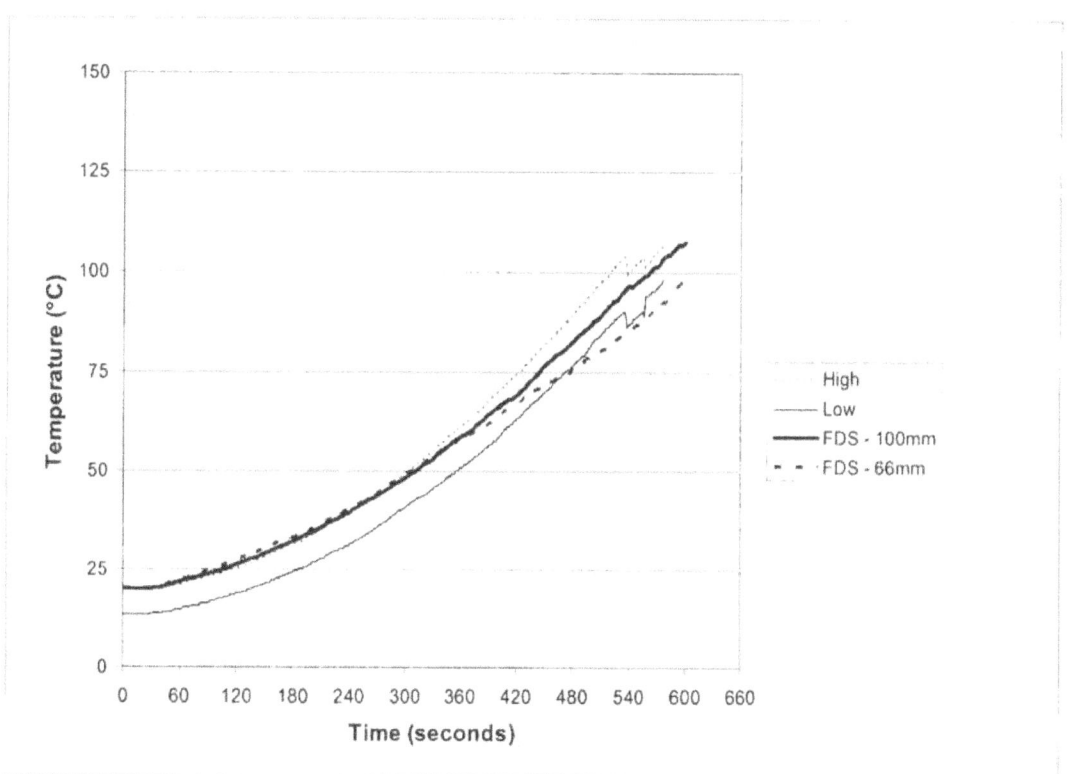

Figure B.47 – Comparison of Predicted and Measured Heat Detector Temperatures, RTI = 164 m$^{1/2}$-s$^{1/2}$, 6.1 m Ceiling Height, Radial Distance = 10.8 m

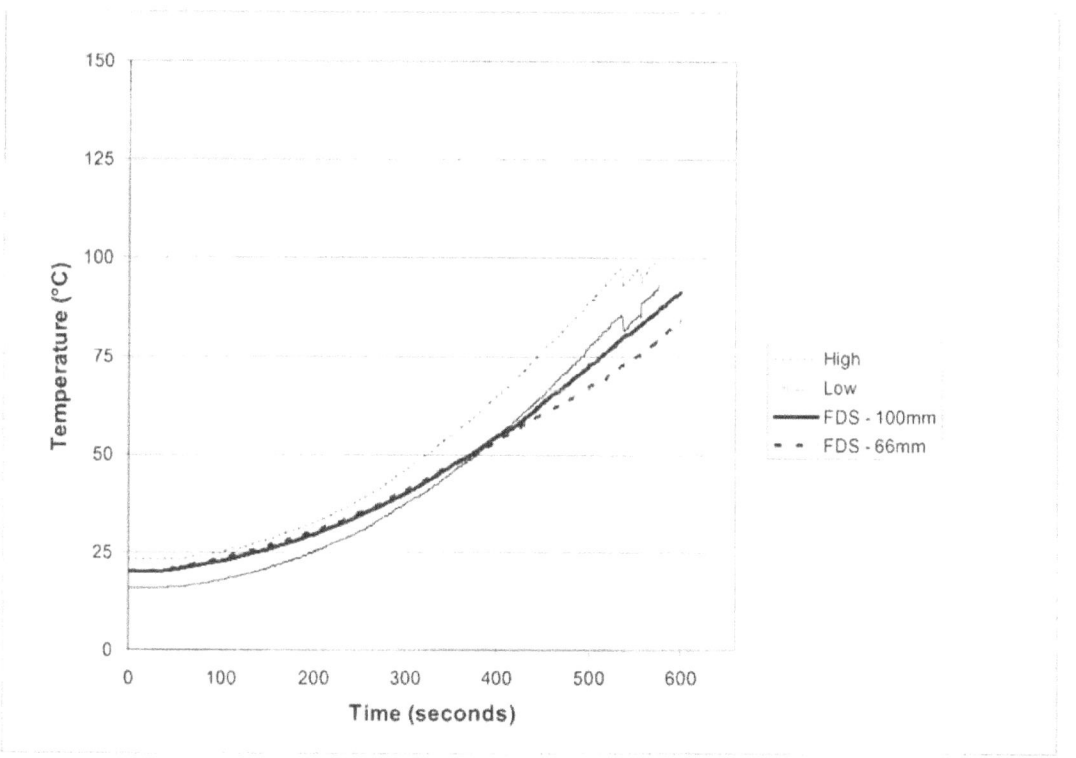

Figure B.48 – Comparison of Predicted and Measured Heat Detector Temperatures, RTI = 287 m$^{1/2}$-s$^{1/2}$, 6.1 m Ceiling Height, Radial Distance = 10.8 m

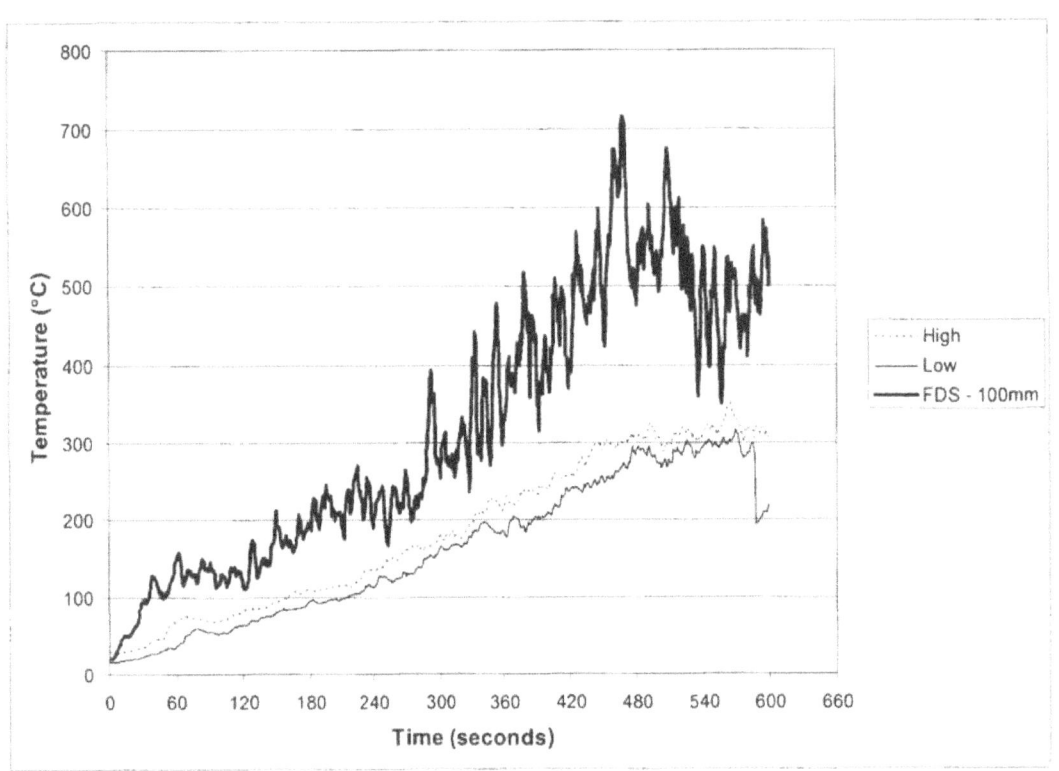

Figure B.49 – Comparison of Predicted and Measured Temperatures, 7.6 m Ceiling Height, Plume Centerline

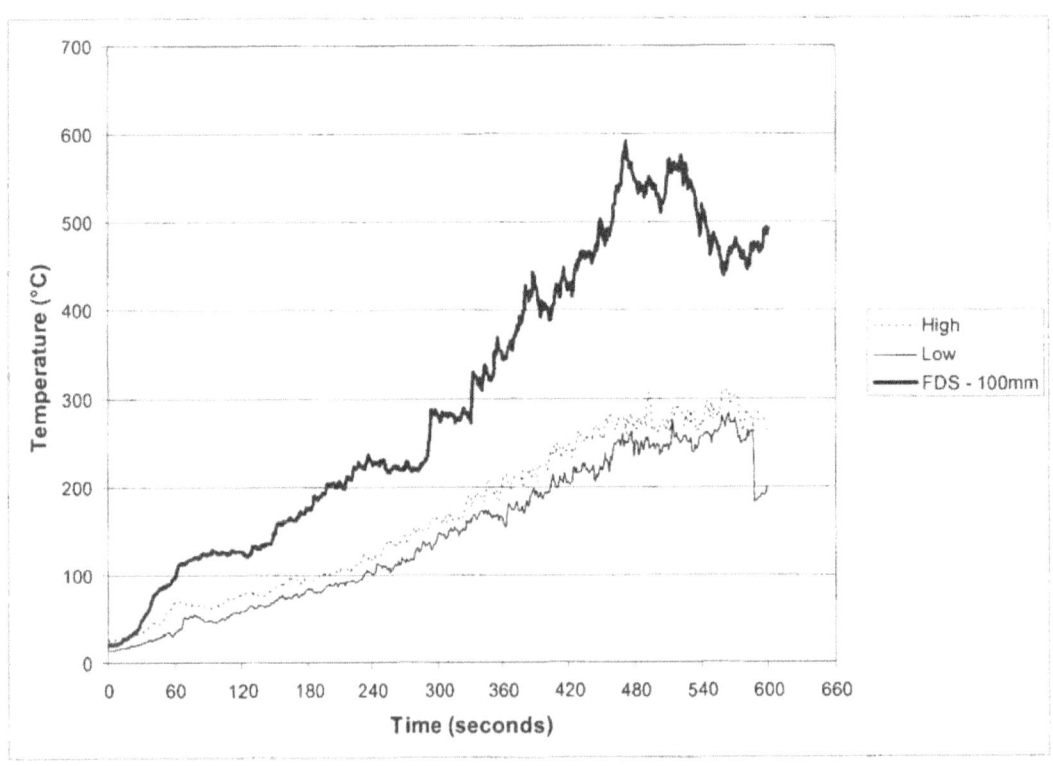

Figure B.50 – Comparison of Predicted and Measured Heat Detector Temperatures, RTI = 32 m$^{1/2}$-s$^{1/2}$, 7.6 m Ceiling Height, Plume Centerline

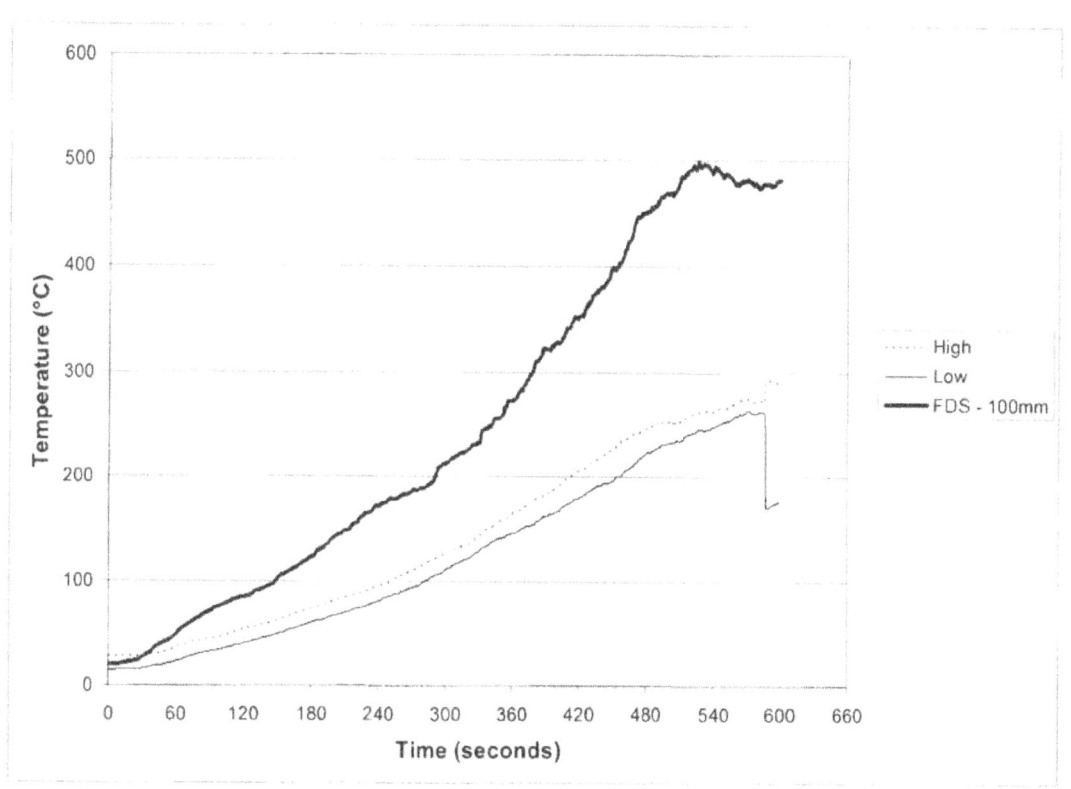

Figure B.51 – Comparison of Predicted and Measured Heat Detector Temperatures, RTI = 164 $m^{1/2}$-$s^{1/2}$, 7.6 m Ceiling Height, Plume Centerline

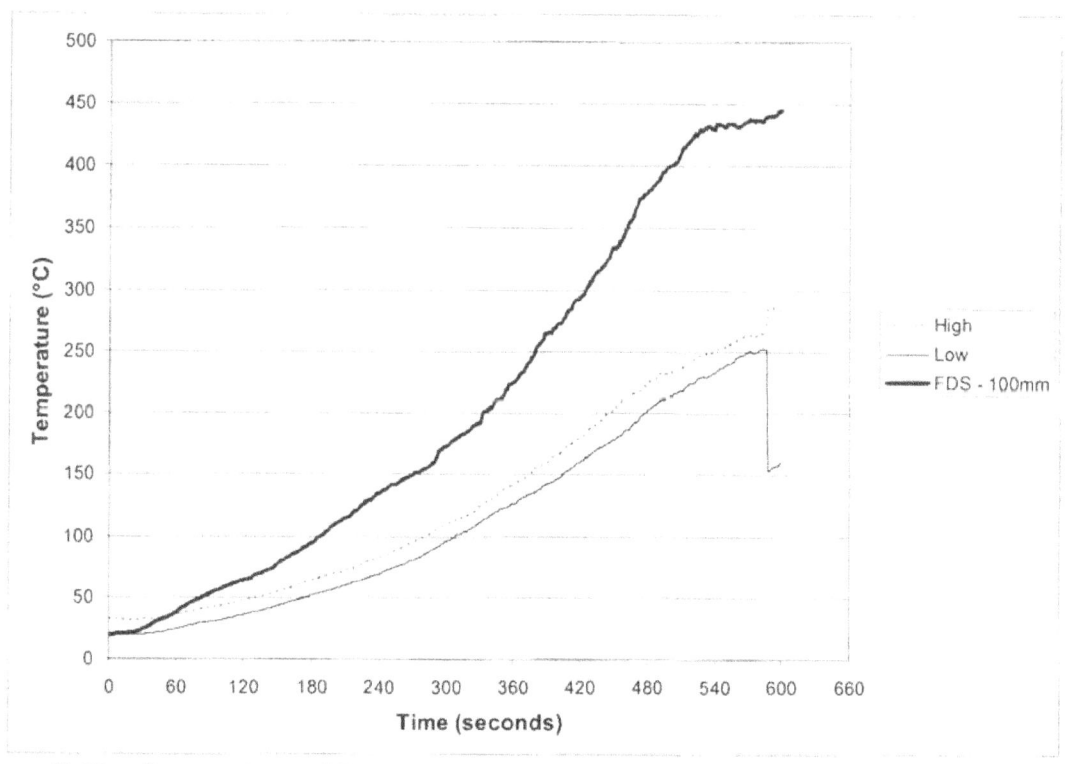

Figure B.52 – Comparison of Predicted and Measured Heat Detector Temperatures, RTI = 287 $m^{1/2}$-$s^{1/2}$, 7.6 m Ceiling Height, Plume Centerline

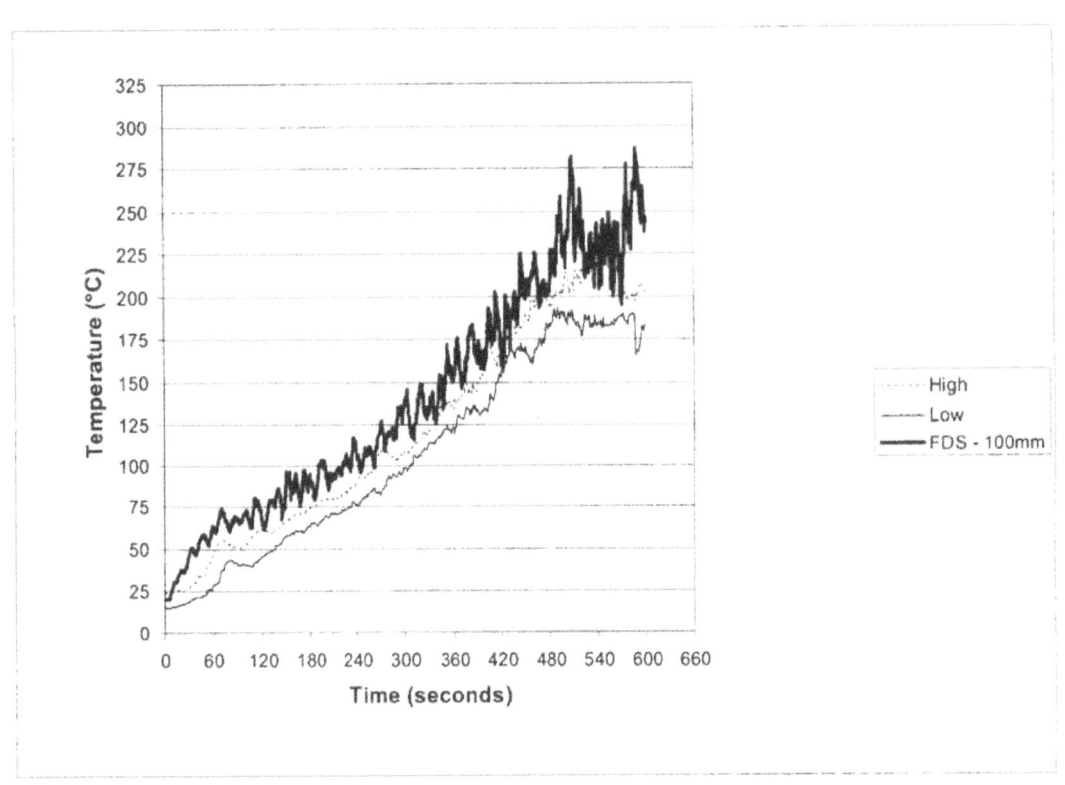

Figure B.53 – Comparison of Predicted and Measured Temperatures, 7.6 m Ceiling Height, Radial Distance = 2.2 m

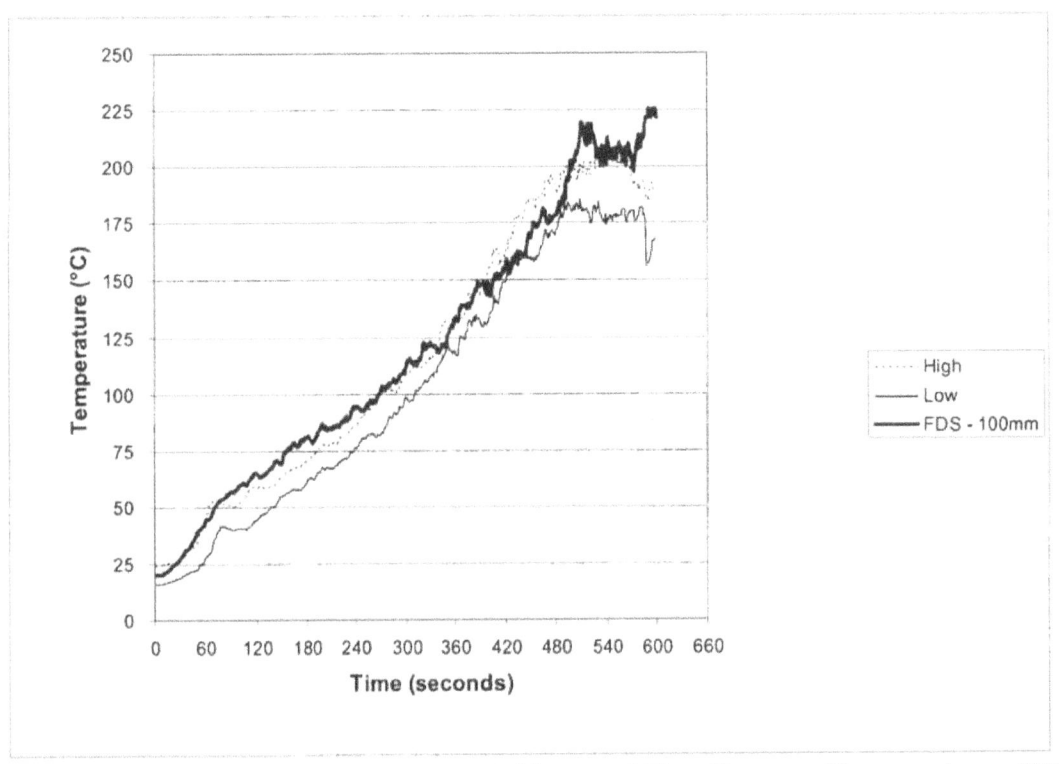

Figure B.54 – Comparison of Predicted and Measured Heat Detector Temperatures, RT1 = 32 $m^{1/2}$-$s^{1/2}$, 7.6 m Ceiling Height, Radial Distance = 2.2 m

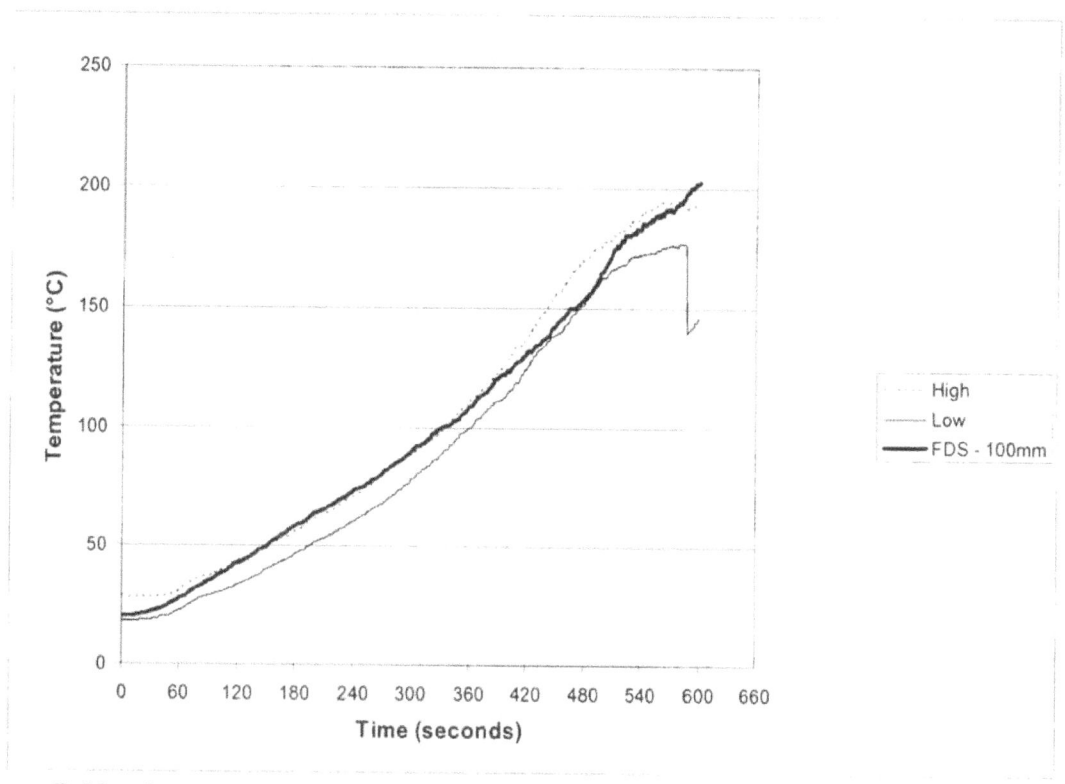

Figure B.55 – Comparison of Predicted and Measured Heat Detector Temperatures, RTI = 164 $m^{1/2}$-$s^{1/2}$, 7.6 m Ceiling Height, Radial Distance = 2.2 m

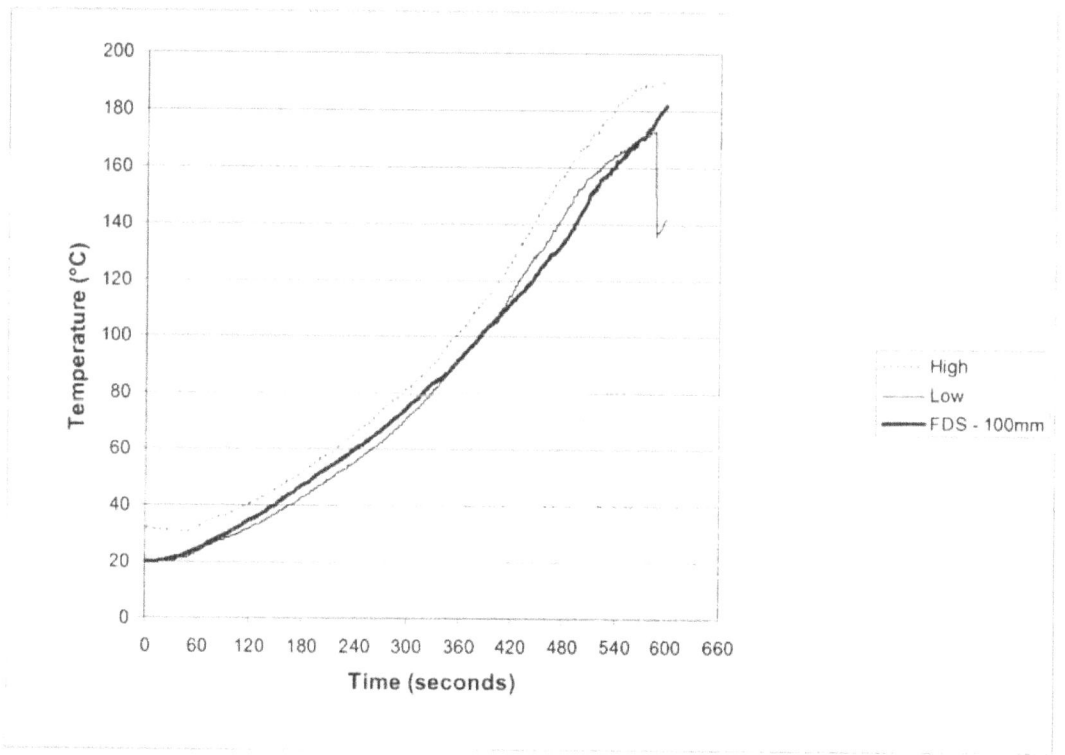

Figure B.56 – Comparison of Predicted and Measured Heat Detector Temperatures, RTI = 287 $m^{1/2}$-$s^{1/2}$, 7.6 m Ceiling Height, Radial Distance = 2.2 m

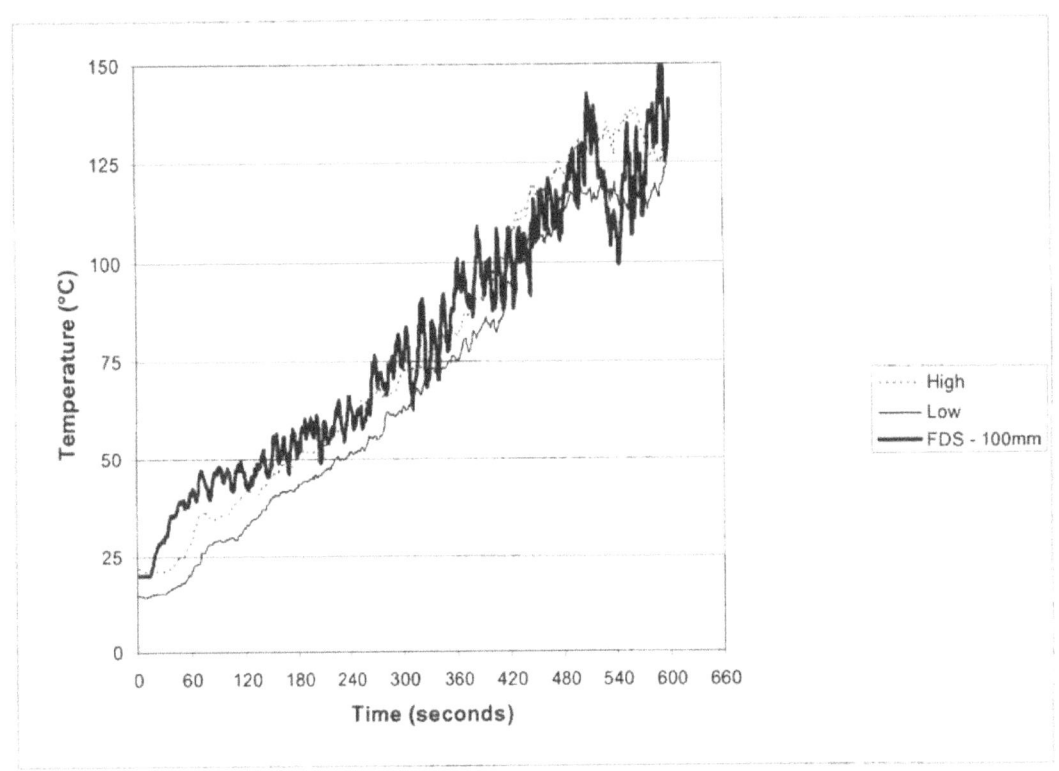

Figure B.57 – Comparison of Predicted and Measured Temperatures, 7.6 m Ceiling Height,
Radial Distance = 6.5 m

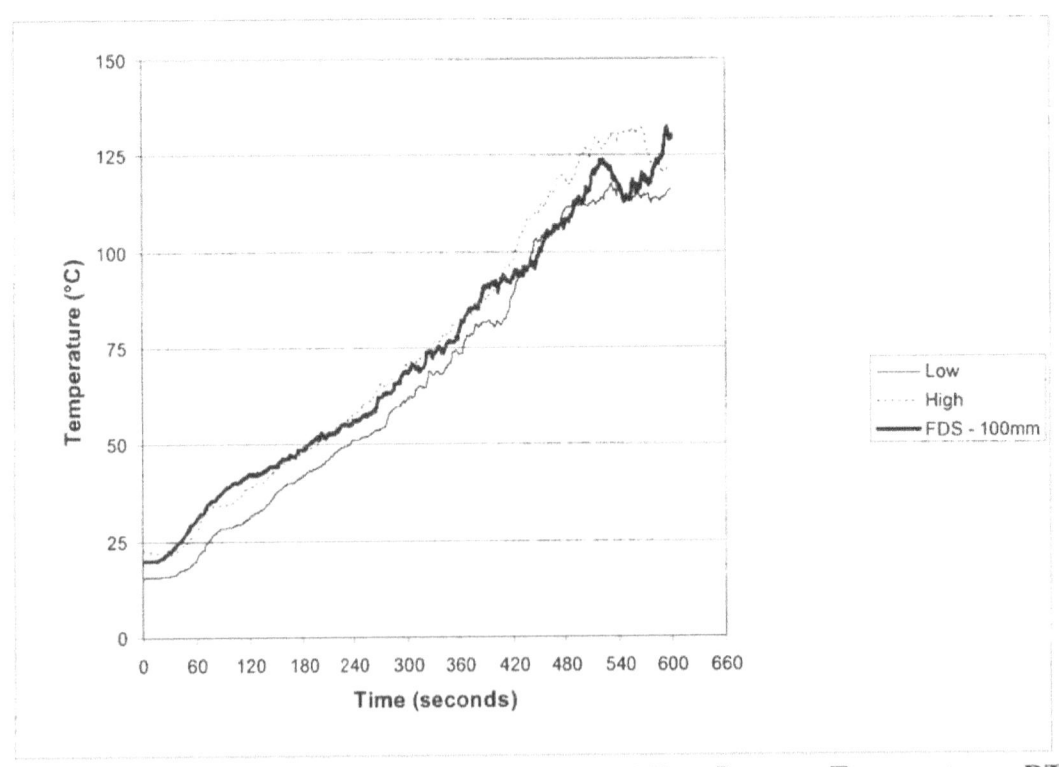

Figure B.58 – Comparison of Predicted and Measured Heat Detector Temperatures, RTI =
32 m$^{1/2}$-s$^{1/2}$, 7.6 m Ceiling Height, Radial Distance = 6.5 m

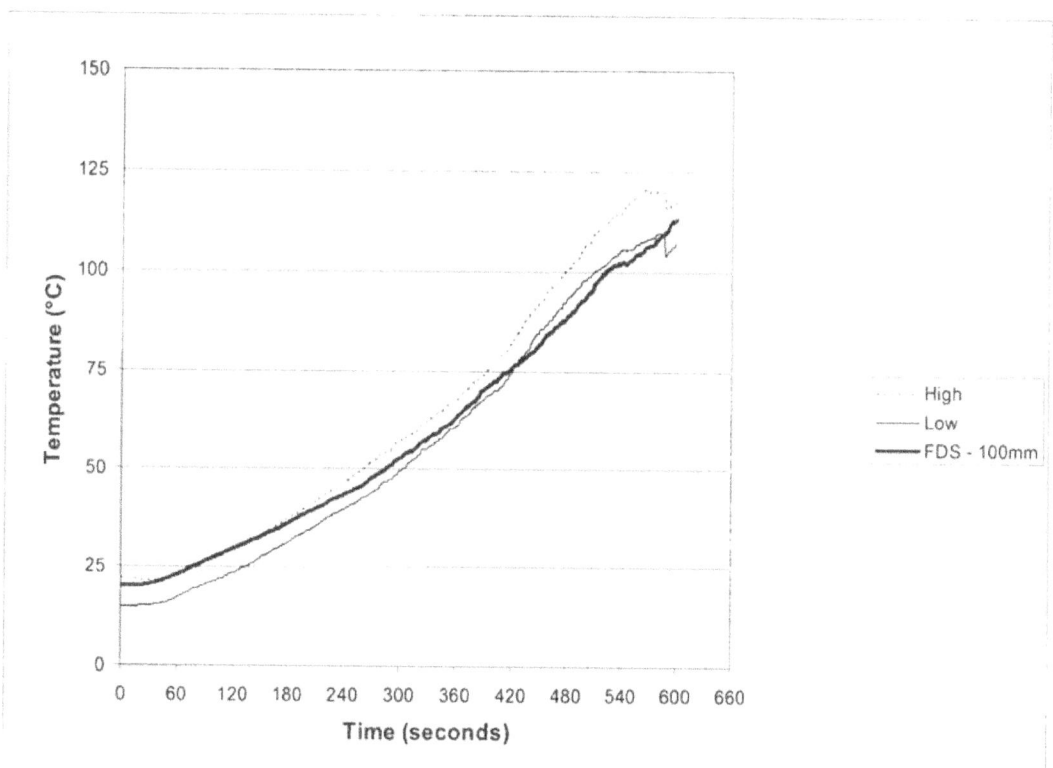

Figure B.59 – Comparison of Predicted and Measured Heat Detector Temperatures, RTI = 164 $m^{1/2}$-$s^{1/2}$, 7.6 m Ceiling Height, Radial Distance = 6.5 m

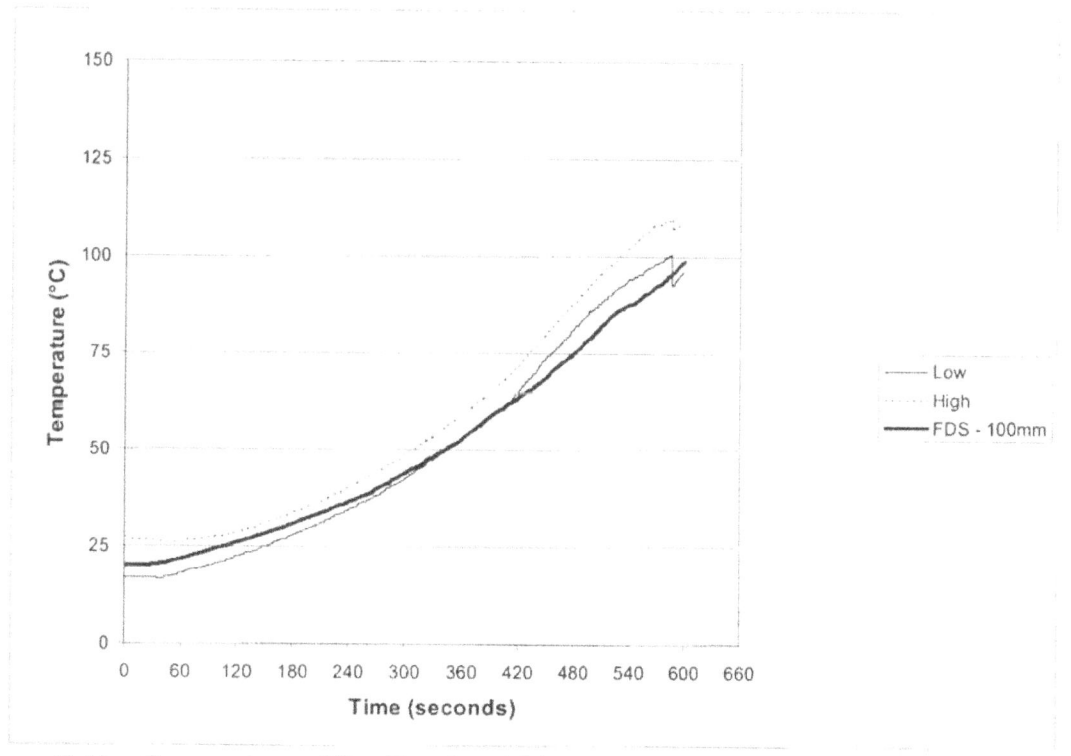

Figure B.60 – Comparison of Predicted and Measured Heat Detector Temperatures, RTI = 287 $m^{1/2}$-$s^{1/2}$, 7.6 m Ceiling Height, Radial Distance = 6.5 m

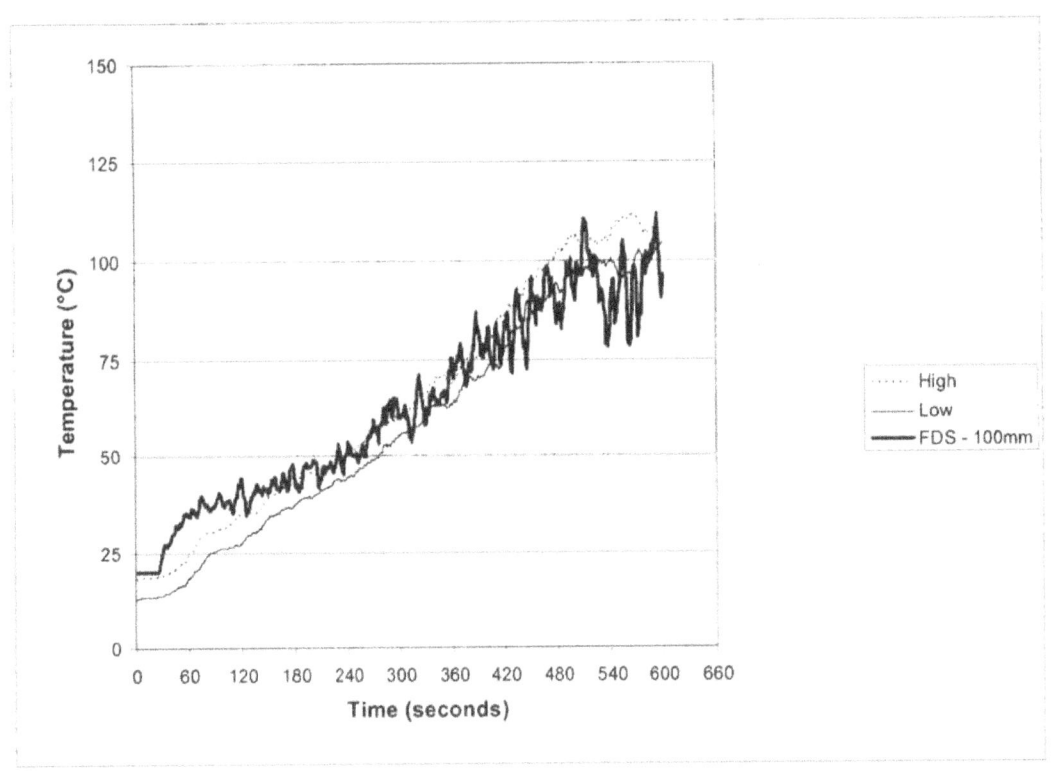

Figure B.61 – Comparison of Predicted and Measured Temperatures, 7.6 m Ceiling Height, Radial Distance = 10.8 m

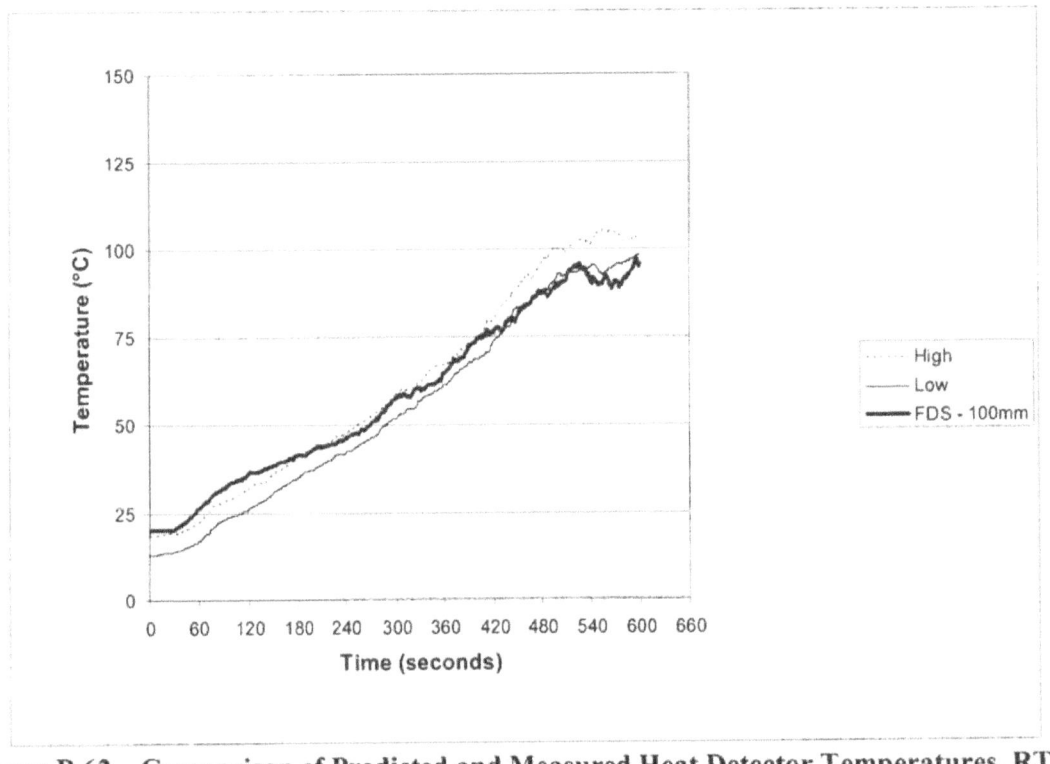

Figure B.62 – Comparison of Predicted and Measured Heat Detector Temperatures, RTI = 32 m$^{1/2}$-s$^{1/2}$, 7.6 m Ceiling Height, Radial Distance = 10.8 m

84

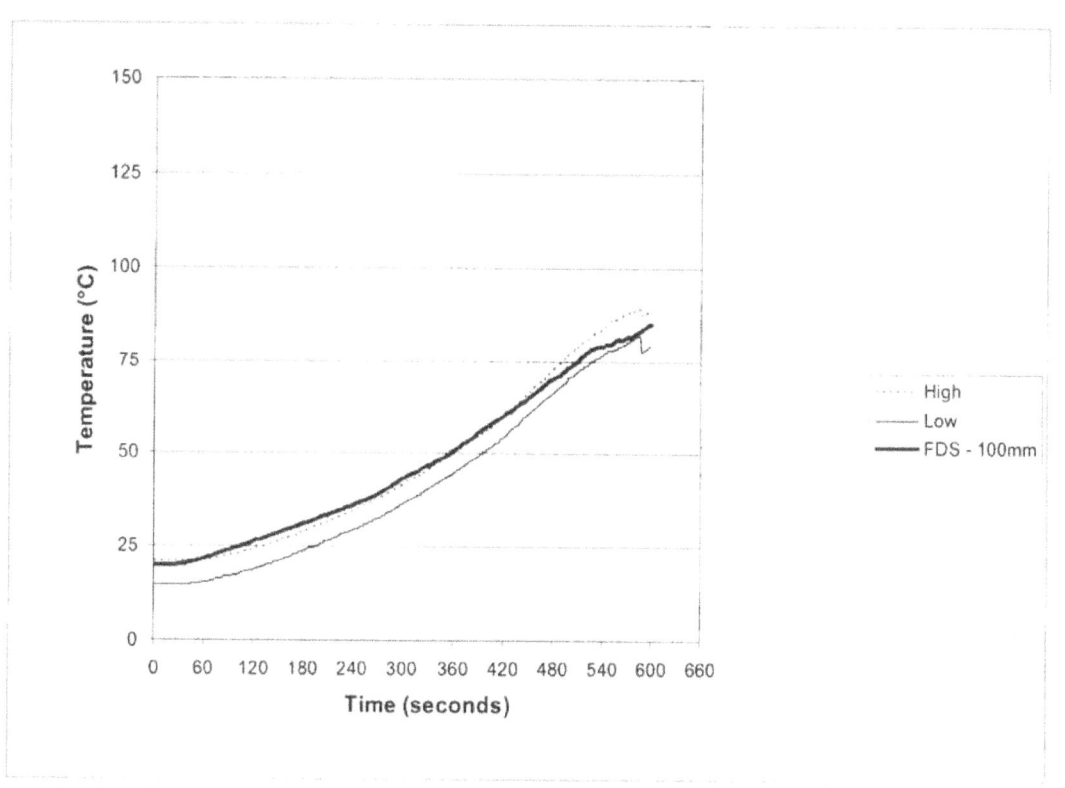

Figure B.63 – Comparison of Predicted and Measured Heat Detector Temperatures, RTI = 164 m$^{1/2}$-s$^{1/2}$, 7.6 m Ceiling Height, Radial Distance = 10.8 m

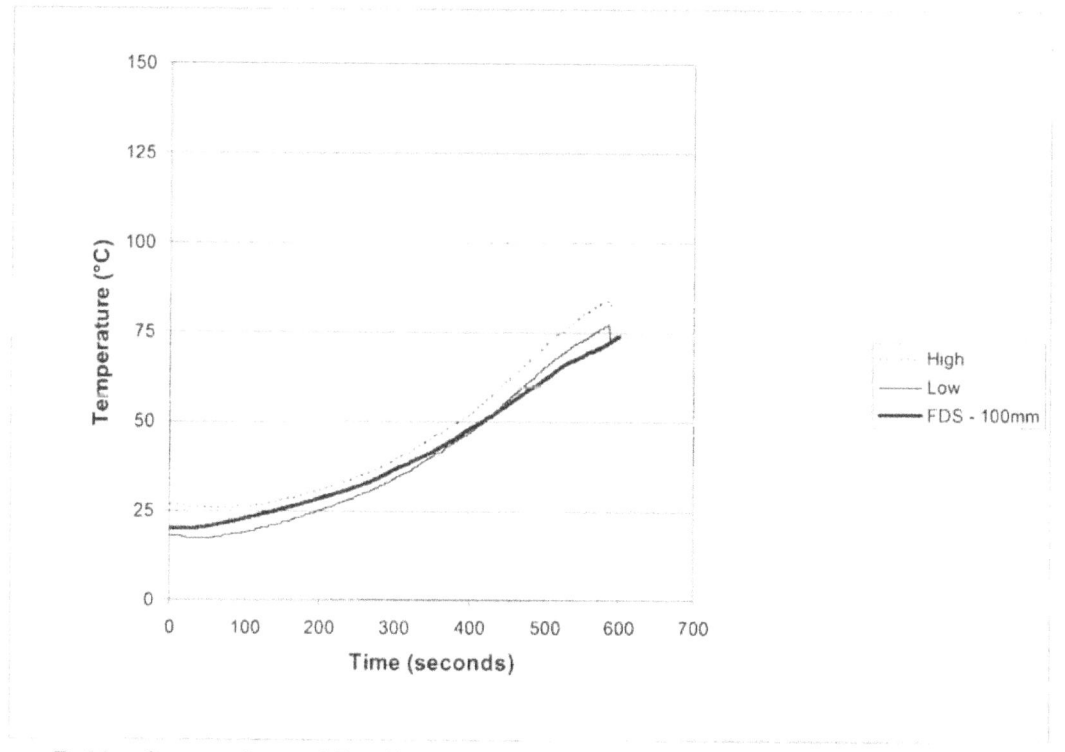

Figure B.64 – Comparison of Predicted and Measured Heat Detector Temperatures, RTI = 287 m$^{1/2}$-s$^{1/2}$, 7.6 m Ceiling Height, Radial Distance = 10.8 m

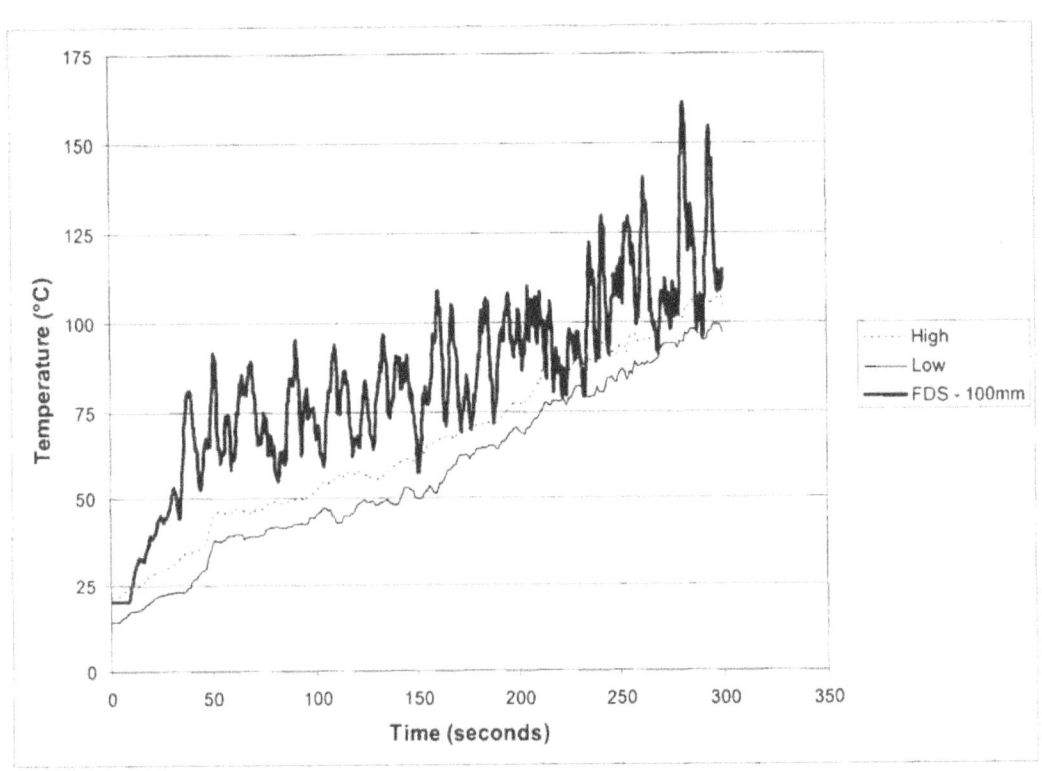

Figure B.65 – Comparison of Predicted and Measured Temperatures, 10.7 m Ceiling Height, Plume Centerline

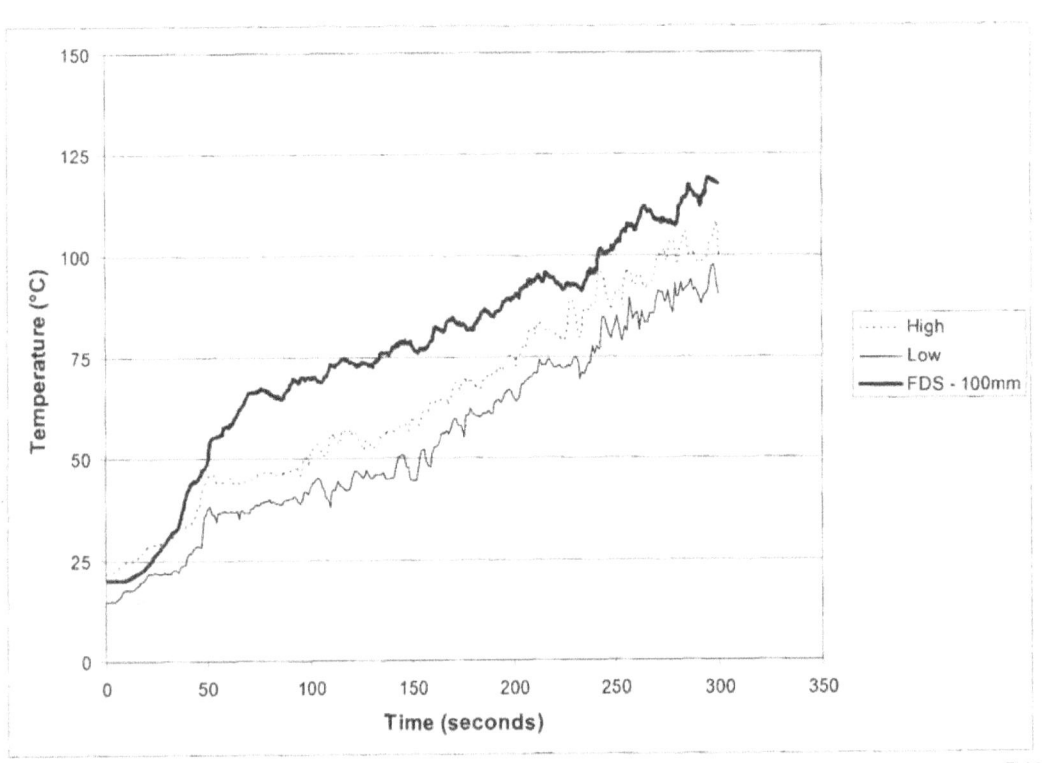

Figure B.66 – Comparison of Predicted and Measured Heat Detector Temperatures, RTI = 32 m$^{1/2}$-s$^{1/2}$, 10.7 m Ceiling Height, Plume Centerline

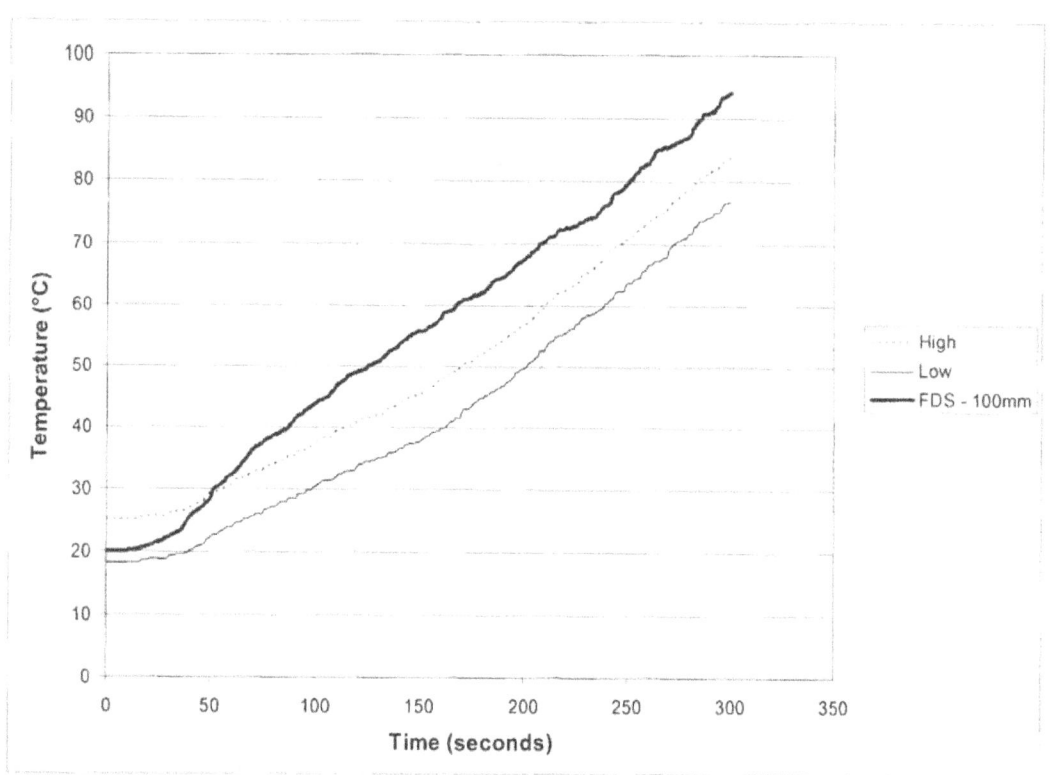

Figure B.67 – Comparison of Predicted and Measured Heat Detector Temperatures, RTI = 164 $m^{1/2}$-$s^{1/2}$, 10.7 m Ceiling Height, Plume Centerline

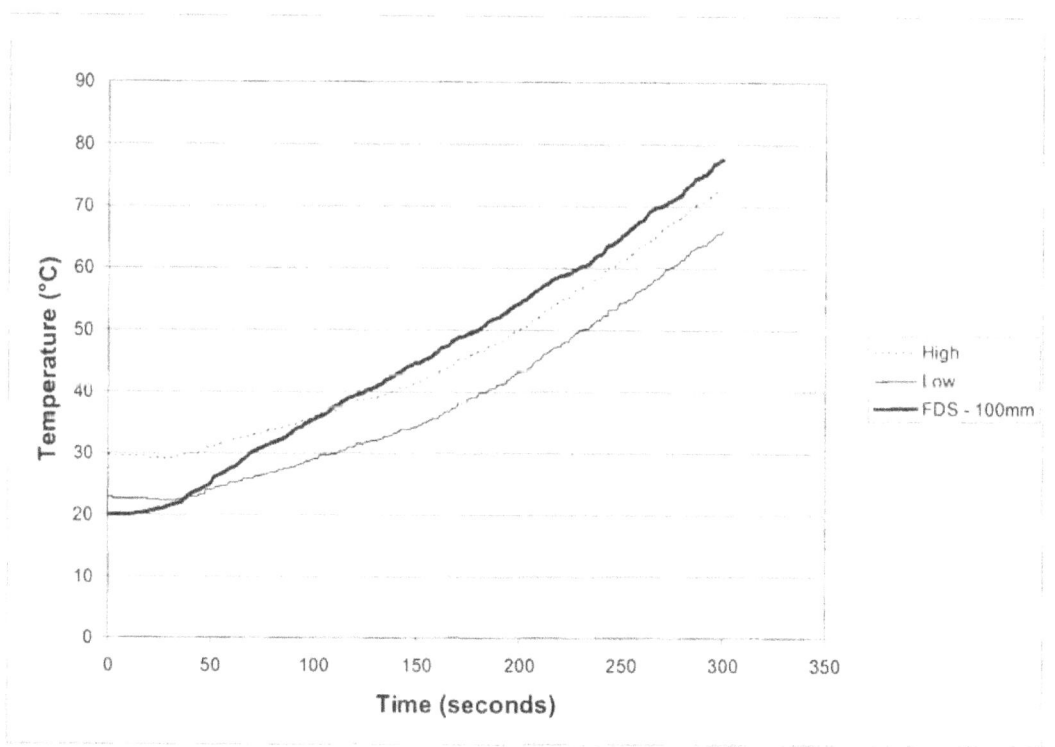

Figure B.68 – Comparison of Predicted and Measured Heat Detector Temperatures, RTI = 287 $m^{1/2}$-$s^{1/2}$, 10.7 m Ceiling Height, Plume Centerline

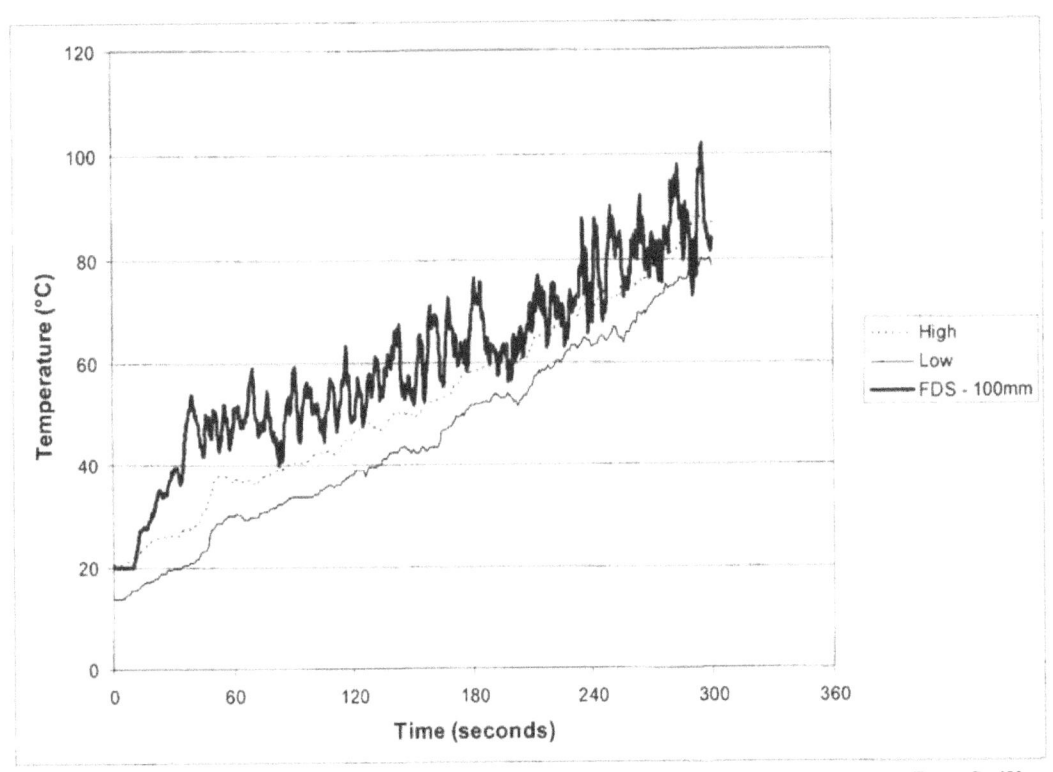

Figure B.69 – Comparison of Predicted and Measured Temperatures, 10.7 m Ceiling Height, Radial Distance = 2.2 m

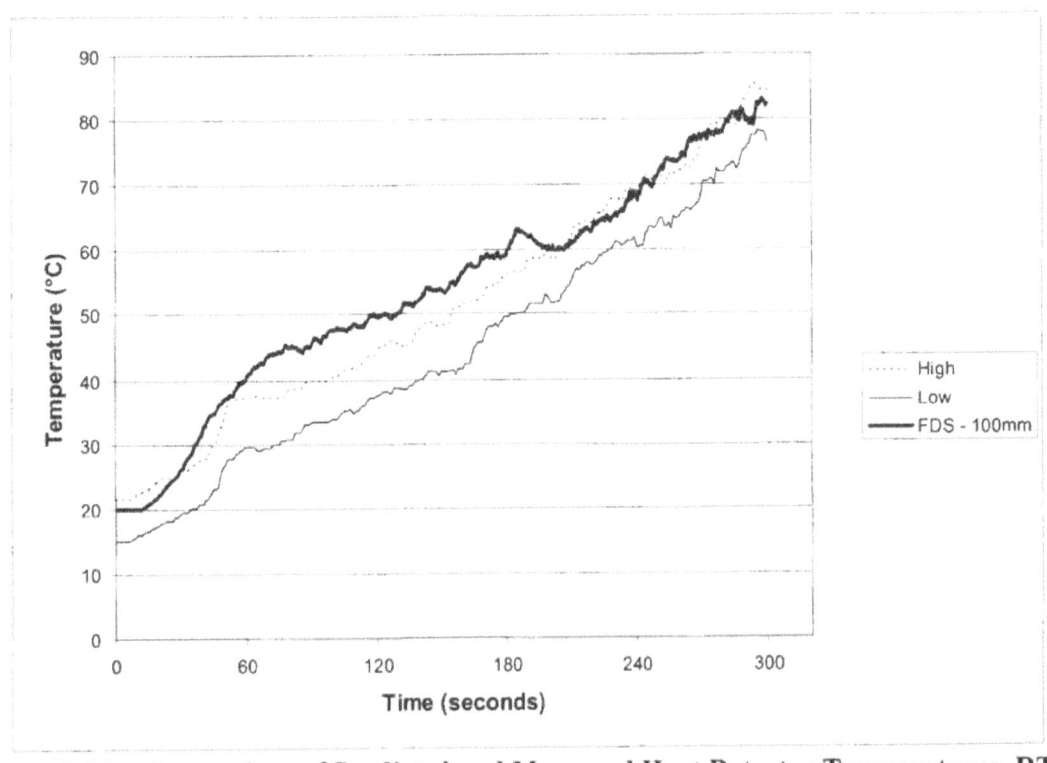

Figure B.70 – Comparison of Predicted and Measured Heat Detector Temperatures, RTI = 32 m$^{1/2}$-s$^{1/2}$, 10.7 m Ceiling Height, Radial Distance = 2.2 m

88

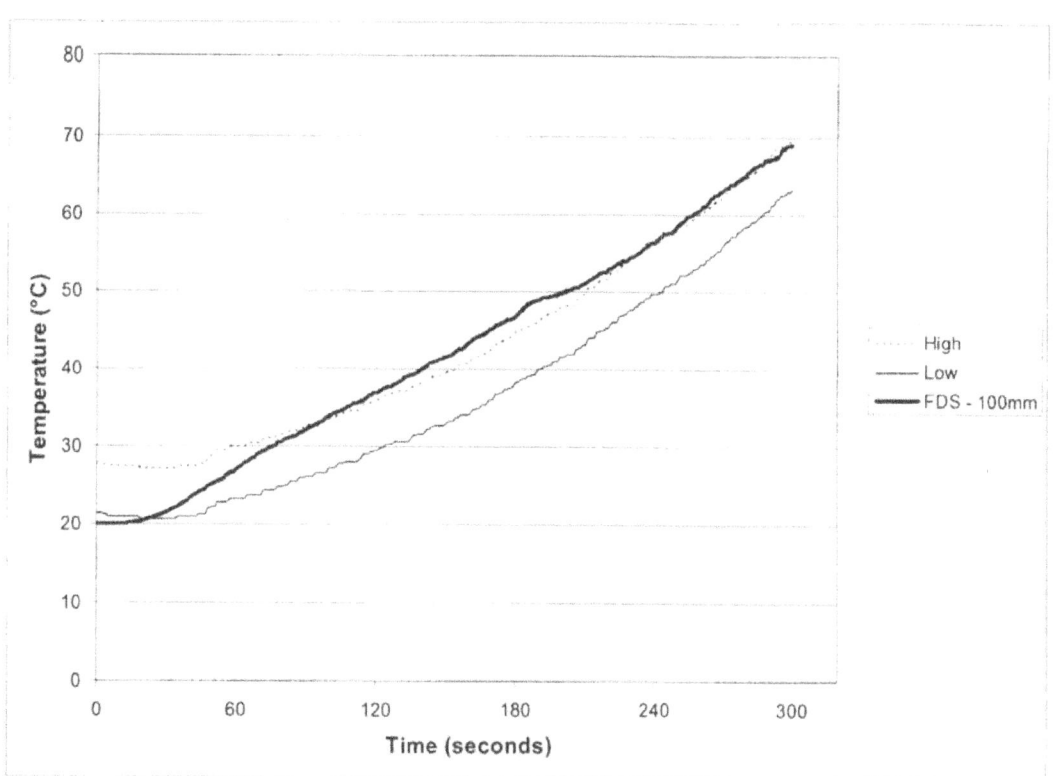

Figure B.71 – Comparison of Predicted and Measured Heat Detector Temperatures, RTI = 164 m$^{1/2}$-s$^{1/2}$, 10.7 m Ceiling Height, Radial Distance = 2.2 m

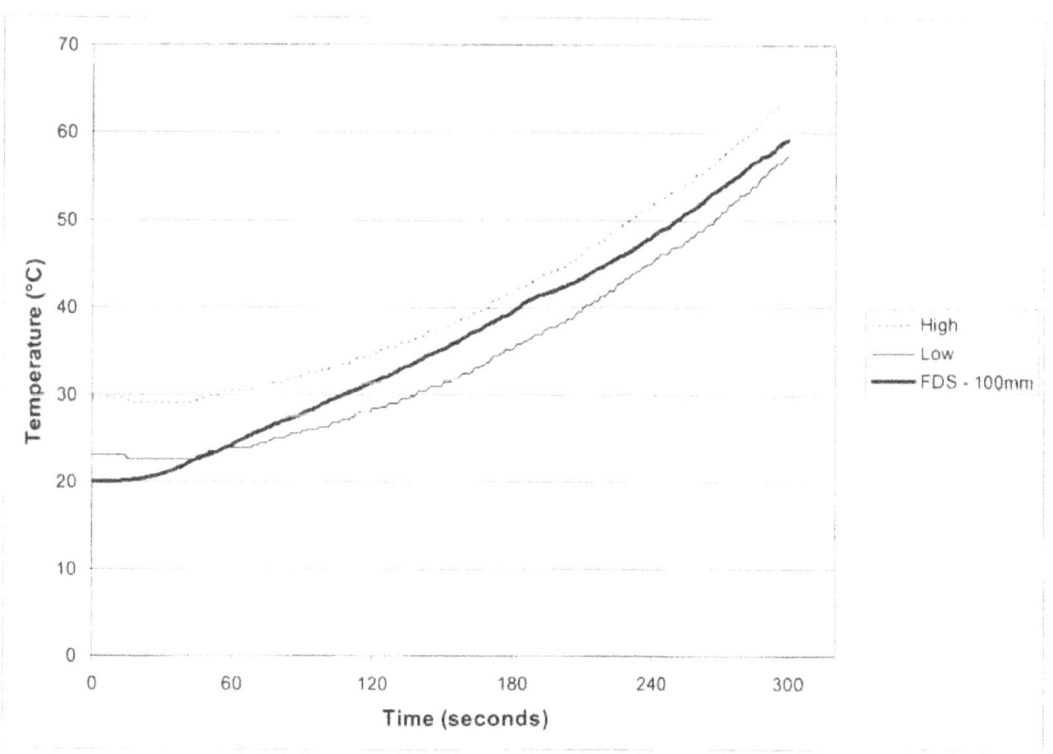

Figure B.72 – Comparison of Predicted and Measured Heat Detector Temperatures, RTI = 287 m$^{1/2}$-s$^{1/2}$, 10.7 m Ceiling Height, Radial Distance = 2.2 m

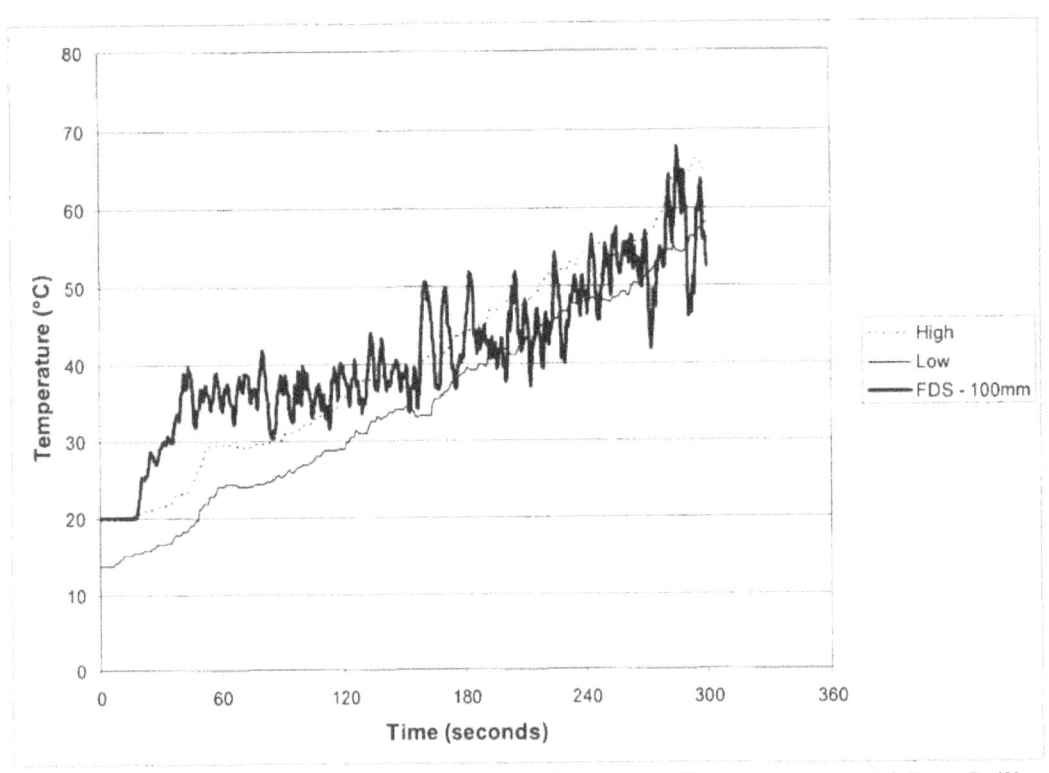

Figure B.73 – Comparison of Predicted and Measured Temperatures, 10.7 m Ceiling Height, Radial Distance = 6.5 m

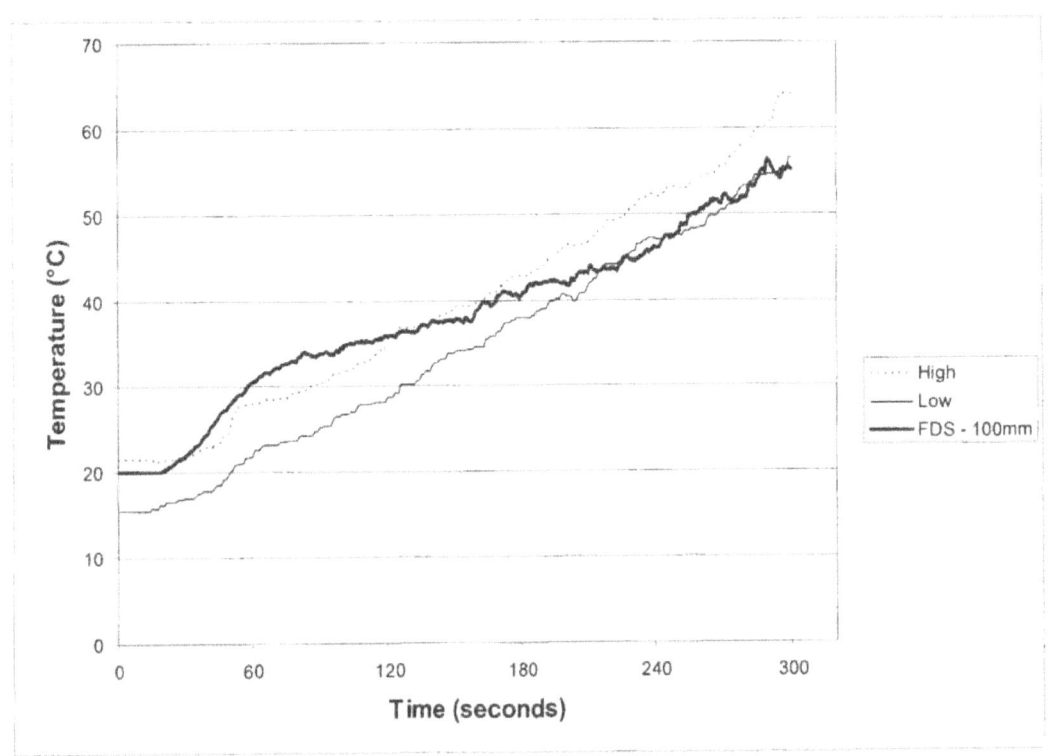

Figure B.74 – Comparison of Predicted and Measured Heat Detector Temperatures, RTI = 32 m$^{1/2}$-s$^{1/2}$, 10.7 m Ceiling Height, Radial Distance = 6.5 m

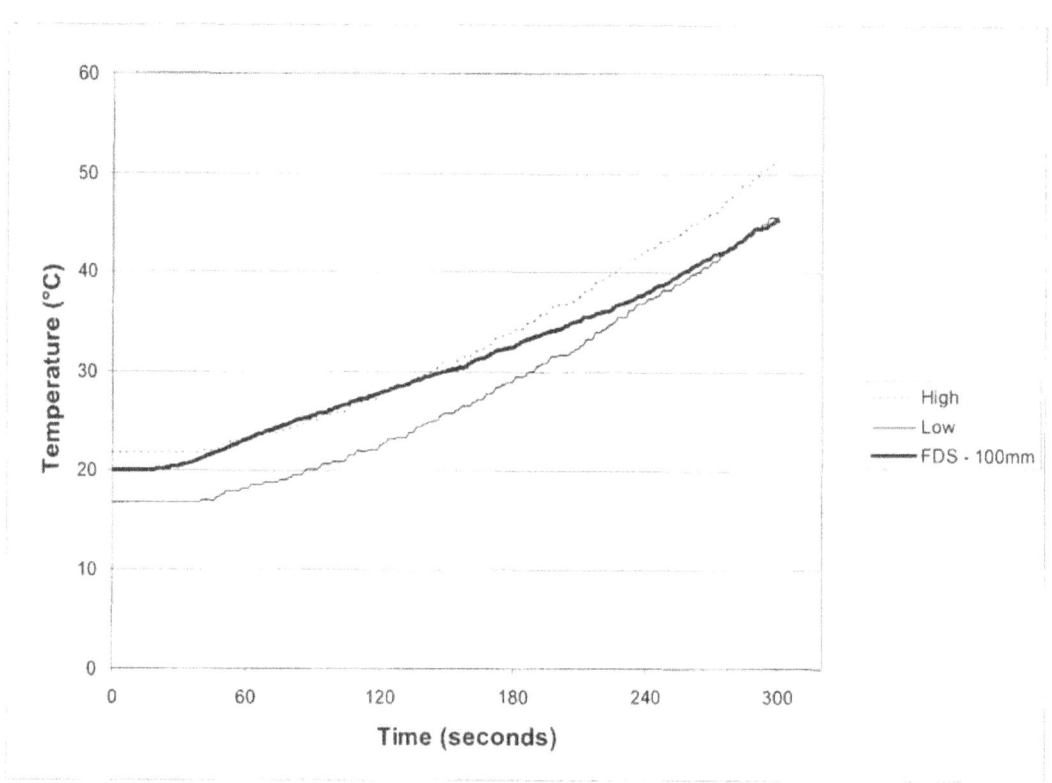

Figure B.75 – Comparison of Predicted and Measured Heat Detector Temperatures, RTI = 164 m$^{1/2}$-s$^{1/2}$, 10.7 m Ceiling Height, Radial Distance = 6.5 m

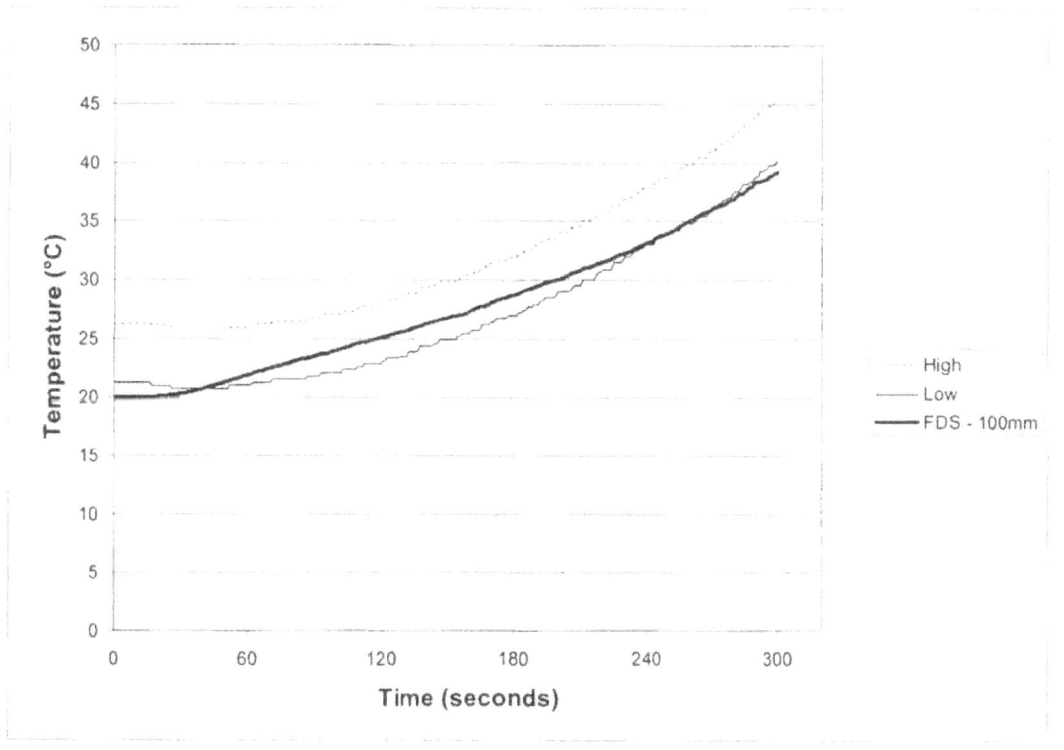

Figure B.76 – Comparison of Predicted and Measured Heat Detector Temperatures, RTI = 287 m$^{1/2}$-s$^{1/2}$, 10.7 m Ceiling Height, Radial Distance = 6.5 m

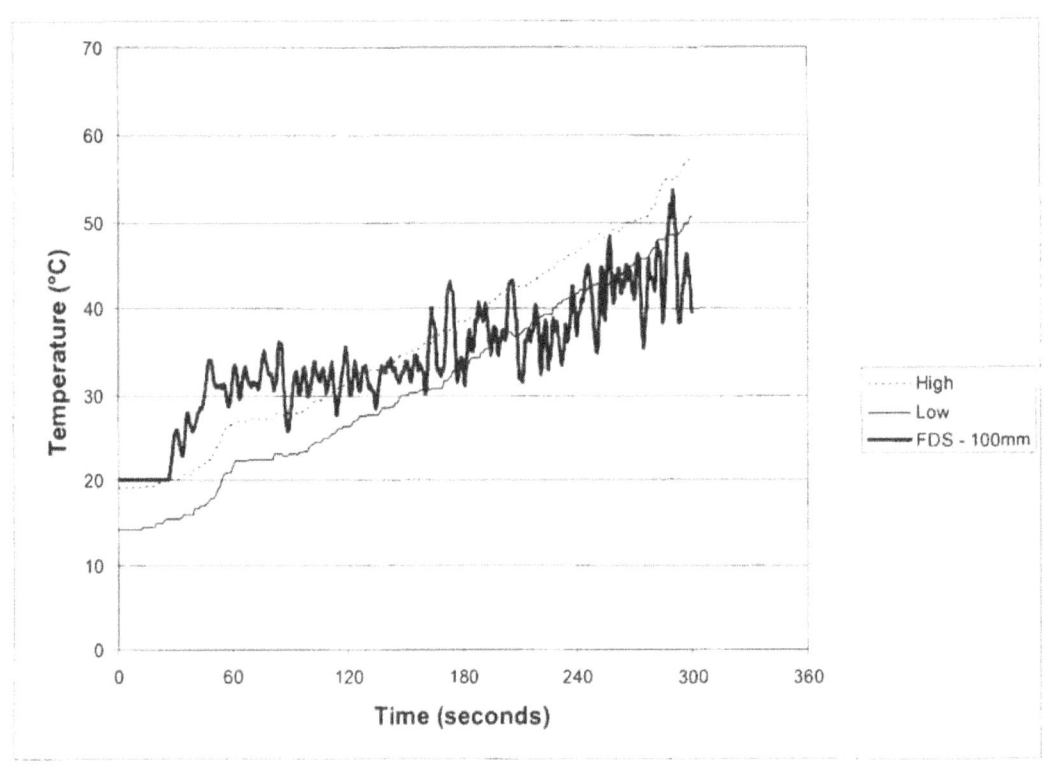

Figure B.77 – Comparison of Predicted and Measured Temperatures, 10.7 m Ceiling Height, Radial Distance = 10.8 m

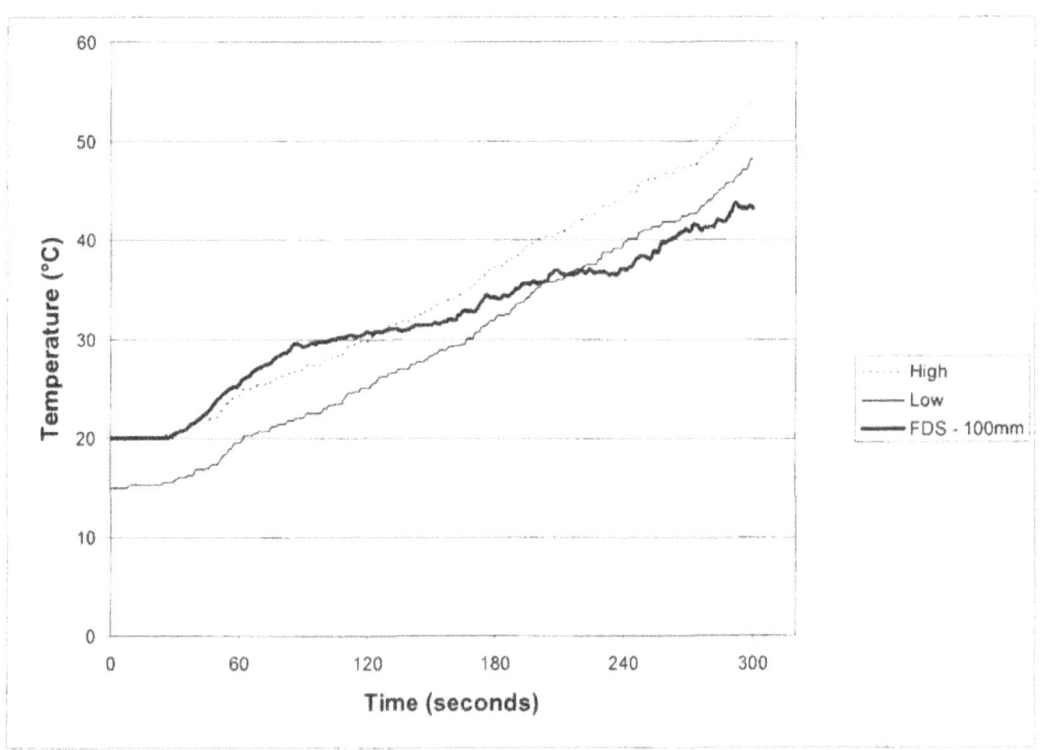

Figure B.78 – Comparison of Predicted and Measured Heat Detector Temperatures, RTI = 32 m$^{1/2}$-s$^{1/2}$, 10.7 m Ceiling Height, Radial Distance = 10.8 m

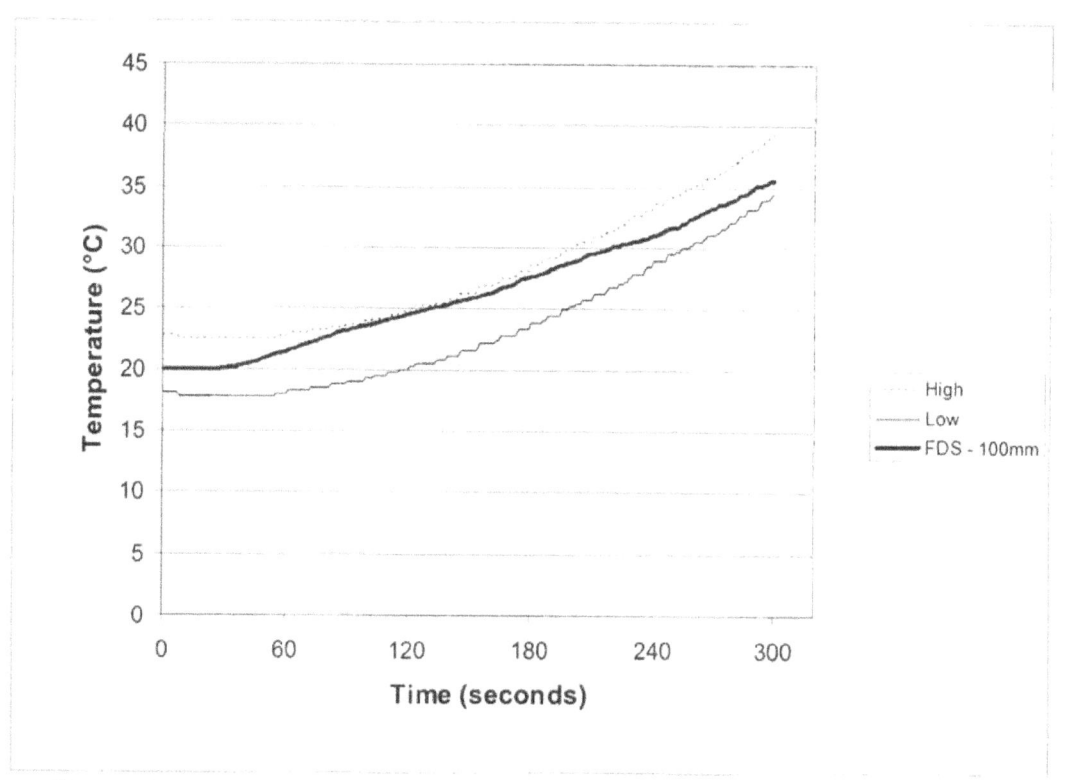

Figure B.79 – Comparison of Predicted and Measured Heat Detector Temperatures, RTI = 164 m$^{1/2}$-s$^{1/2}$, 10.7 m Ceiling Height, Radial Distance = 10.8 m

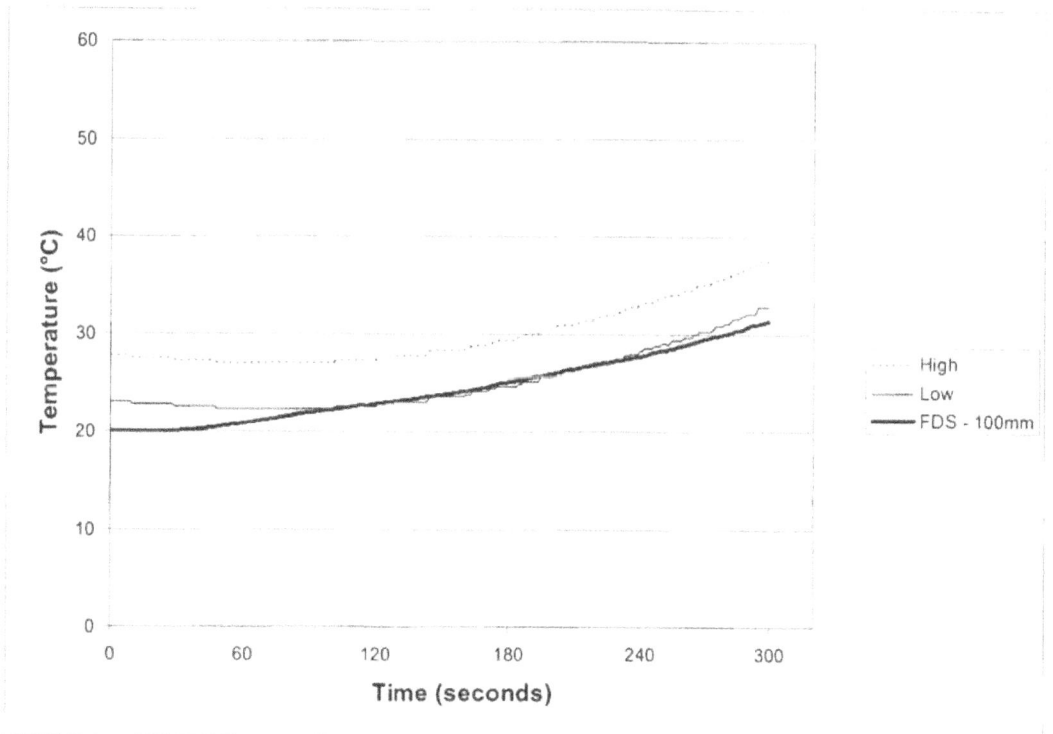

Figure B.80 – Comparison of Predicted and Measured Heat Detector Temperatures, RTI = 287 m$^{1/2}$-s$^{1/2}$, 10.7 m Ceiling Height, Radial Distance = 10.8 m

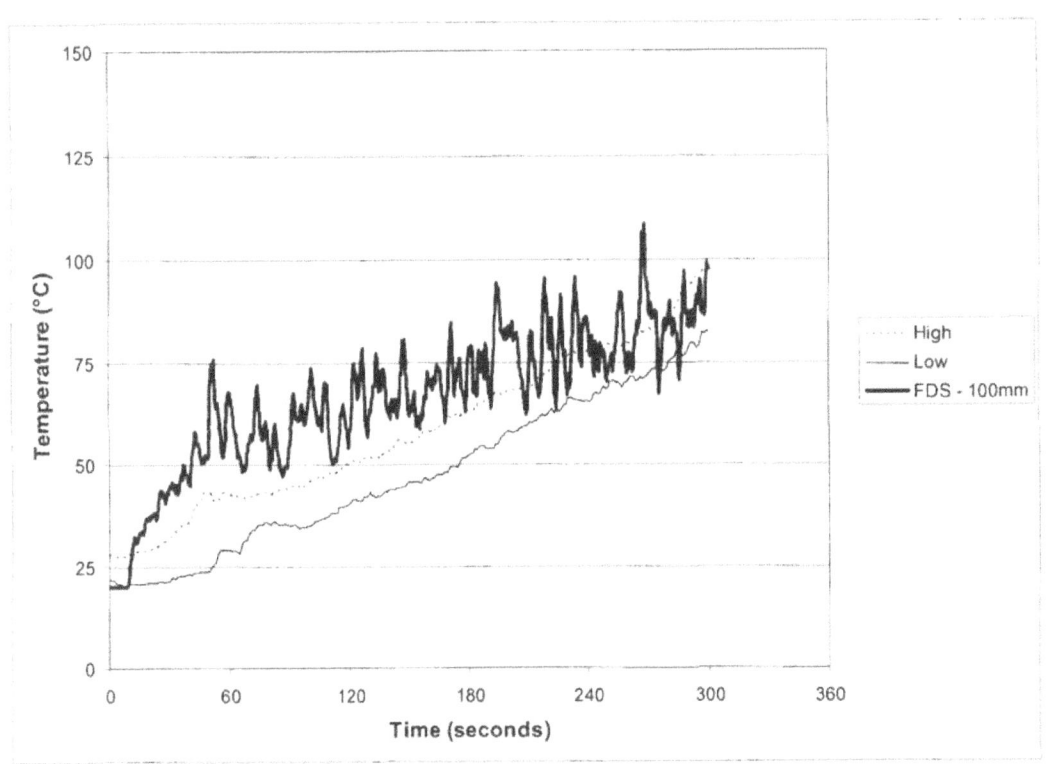

Figure B.81 – Comparison of Predicted and Measured Temperatures, 10.7 m Ceiling Height, Plume Centerline

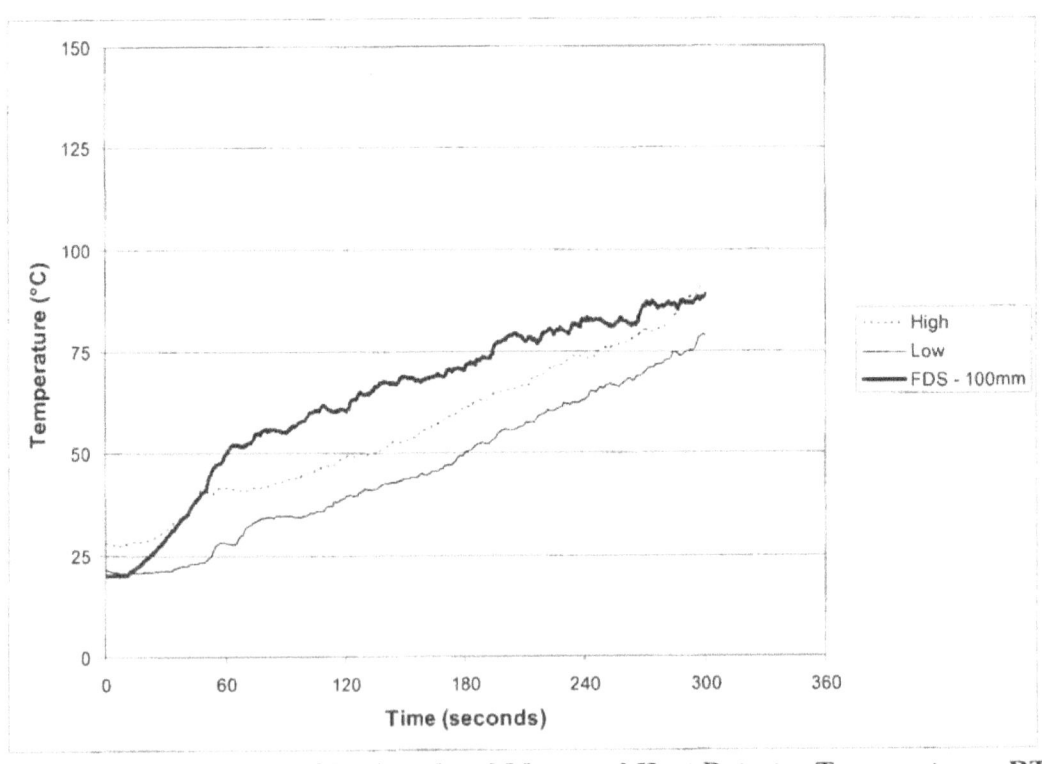

Figure B.82 – Comparison of Predicted and Measured Heat Detector Temperatures, RTI = 32 m$^{1/2}$-s$^{1/2}$, 10.7 m Ceiling Height, Plume Centerline

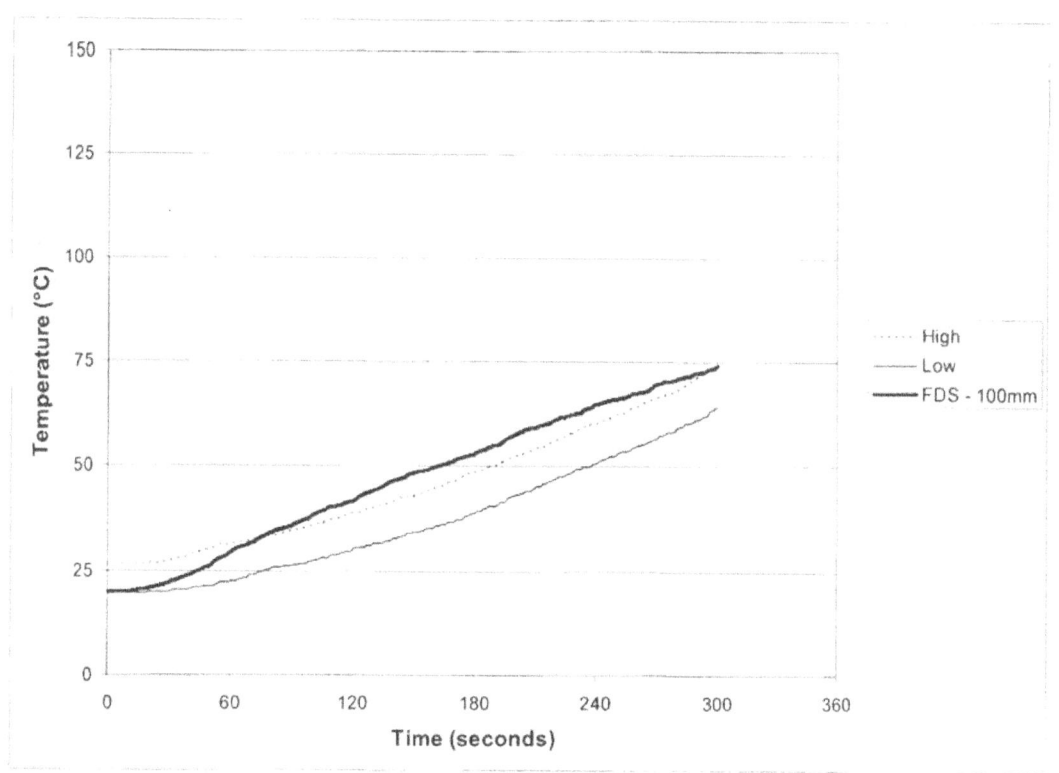

Figure B.83 – Comparison of Predicted and Measured Heat Detector Temperatures, RTI = 164 m$^{1/2}$-s$^{1/2}$, 10.7 m Ceiling Height, Plume Centerline

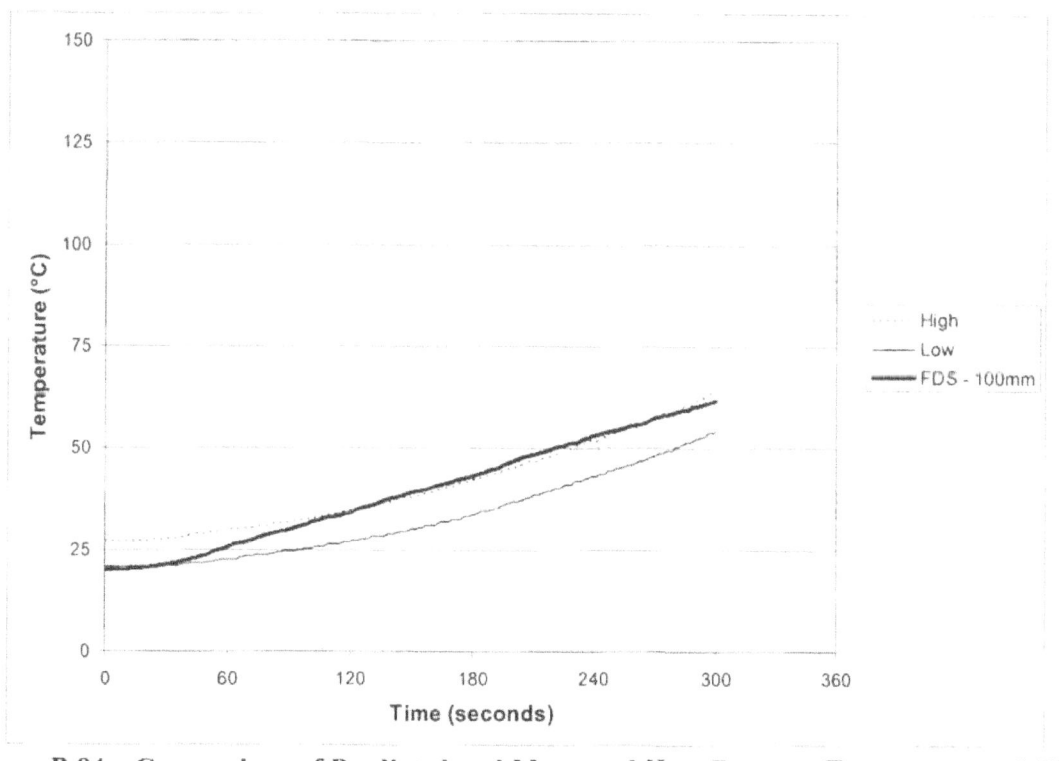

Figure B.84 – Comparison of Predicted and Measured Heat Detector Temperatures, RTI = 287 m$^{1/2}$-s$^{1/2}$, 10.7 m Ceiling Height, Plume Centerline

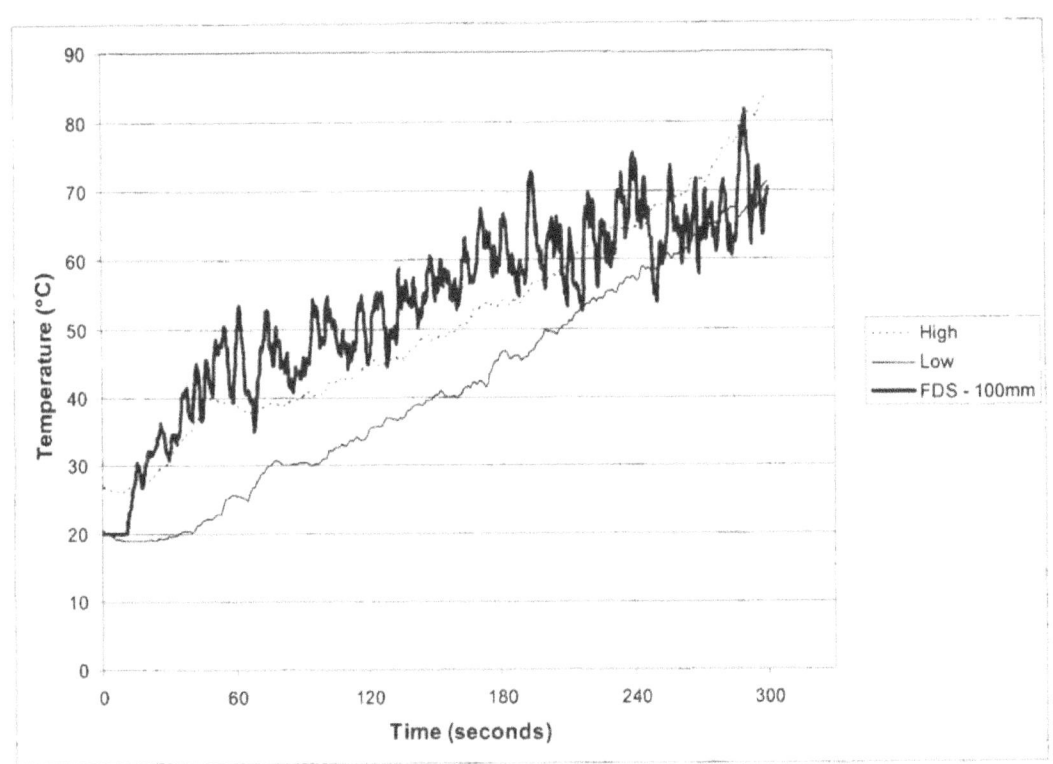

Figure B.85 – Comparison of Predicted and Measured Temperatures, 10.7 m Ceiling Height, Radial Distance = 2.2 m

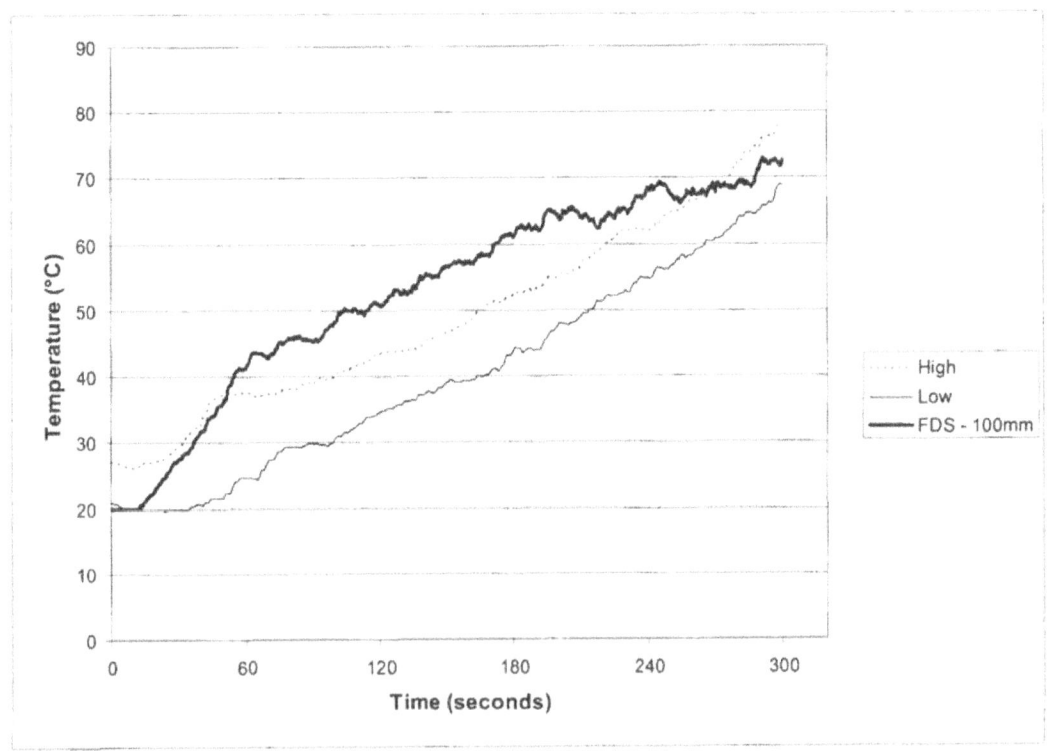

Figure B.86 – Comparison of Predicted and Measured Heat Detector Temperatures, RTI = 32 m$^{1/2}$-s$^{1/2}$, 10.7 m Ceiling Height, Radial Distance = 2.2 m

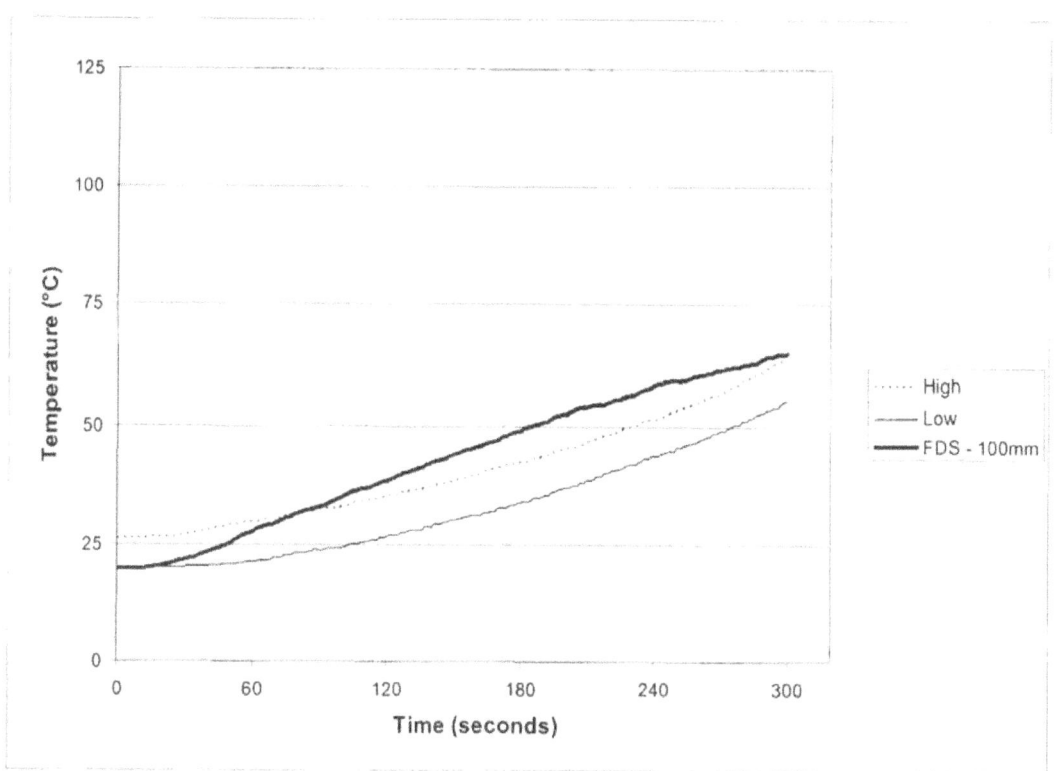

Figure B.87 – Comparison of Predicted and Measured Heat Detector Temperatures, RTI = 164 $m^{1/2}$-$s^{1/2}$, 10.7 m Ceiling Height, Radial Distance = 2.2 m

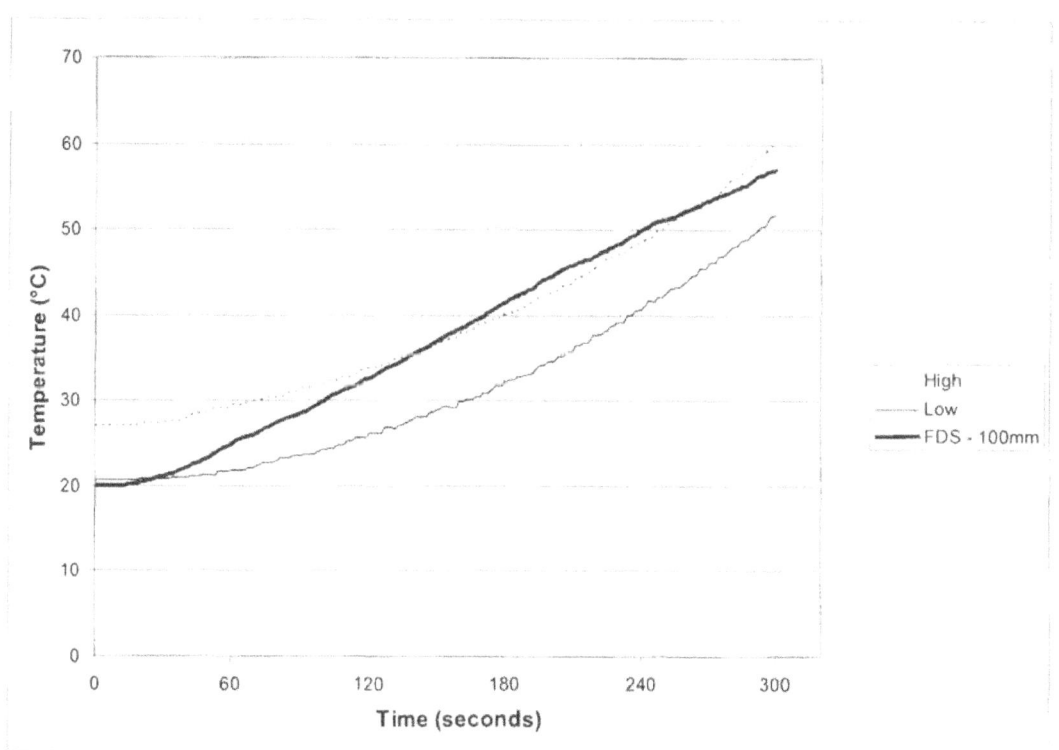

Figure B.88 – Comparison of Predicted and Measured Heat Detector Temperatures, RTI = 287 $m^{1/2}$-$s^{1/2}$, 10.7 m Ceiling Height, Radial Distance = 2.2 m

97

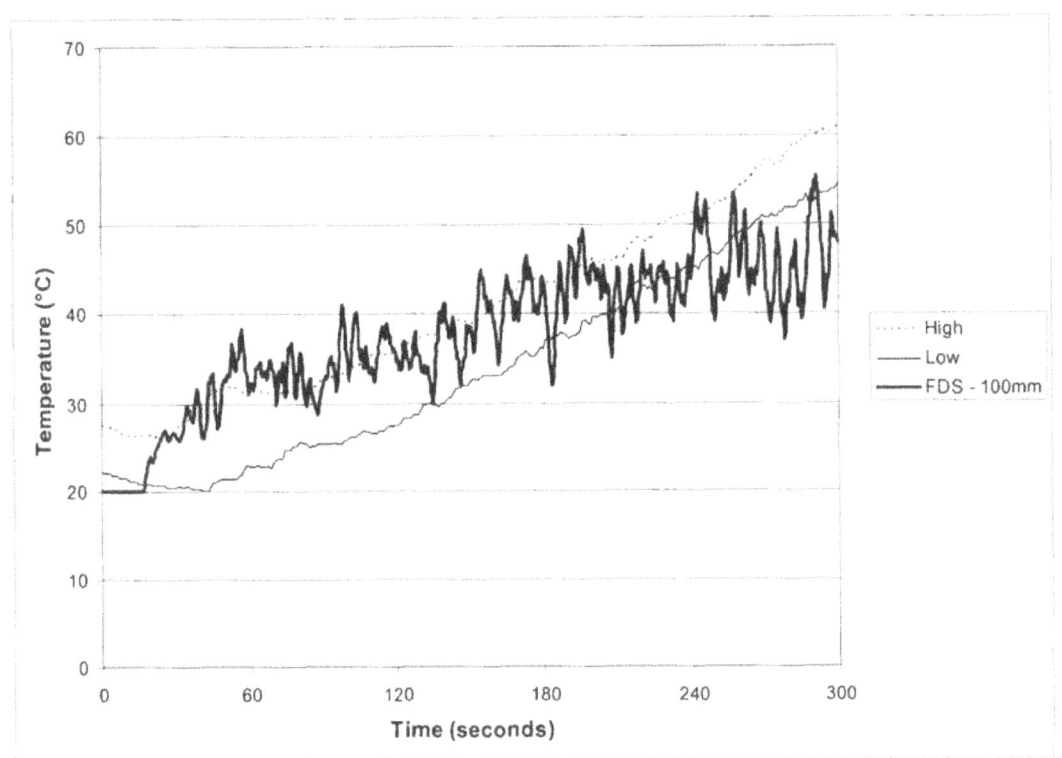

Figure B.89 – Comparison of Predicted and Measured Temperatures, 10.7 m Ceiling Height, Radial Distance = 6.5 m

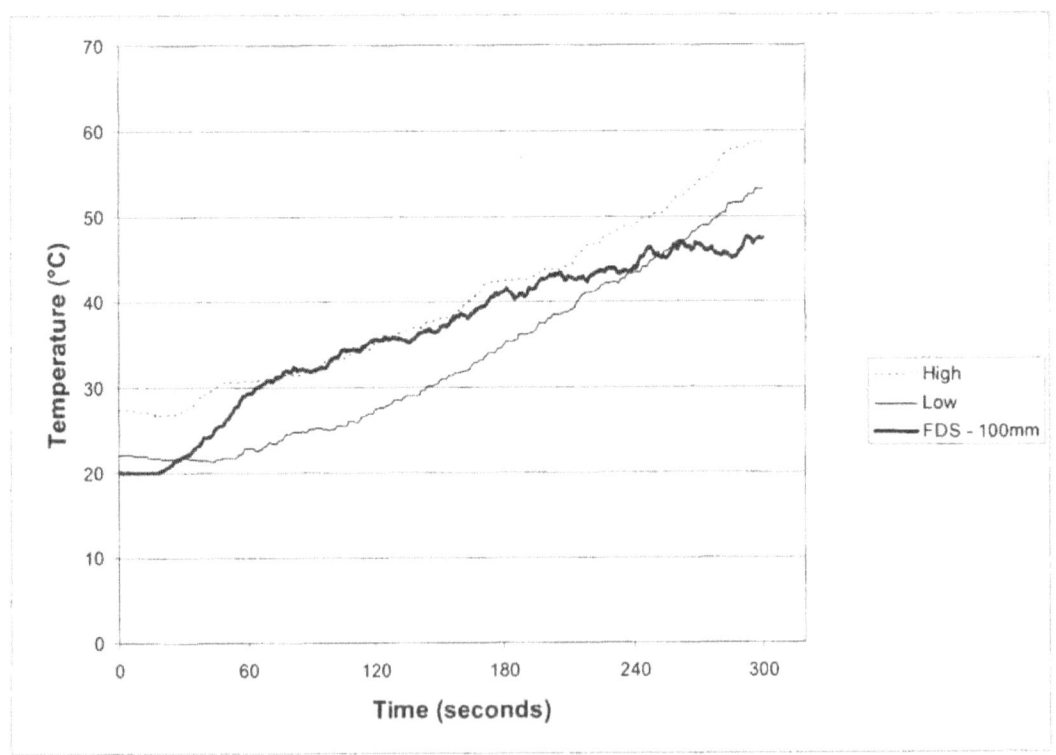

Figure B.90 – Comparison of Predicted and Measured Heat Detector Temperatures, RTI = 32 $m^{1/2}$-$s^{1/2}$, 10.7 m Ceiling Height, Radial Distance = 6.5 m

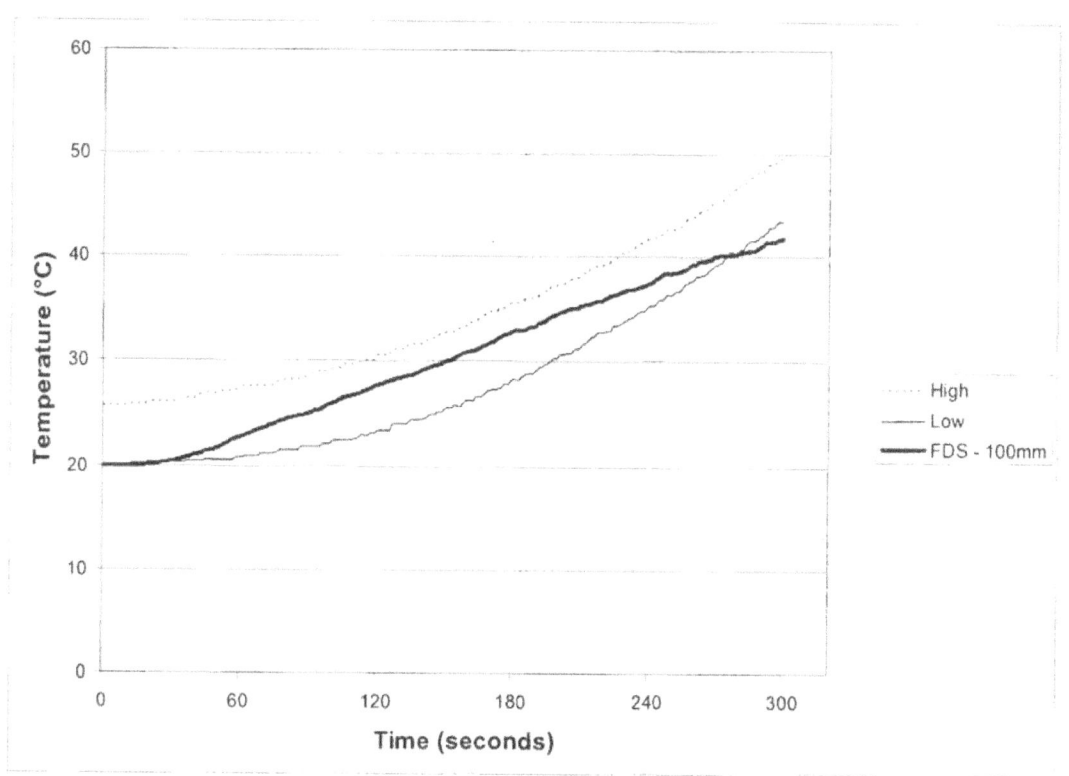

Figure B.91 – Comparison of Predicted and Measured Heat Detector Temperatures, RTI = 164 m$^{1/2}$-s$^{1/2}$, 10.7 m Ceiling Height, Radial Distance = 6.5 m

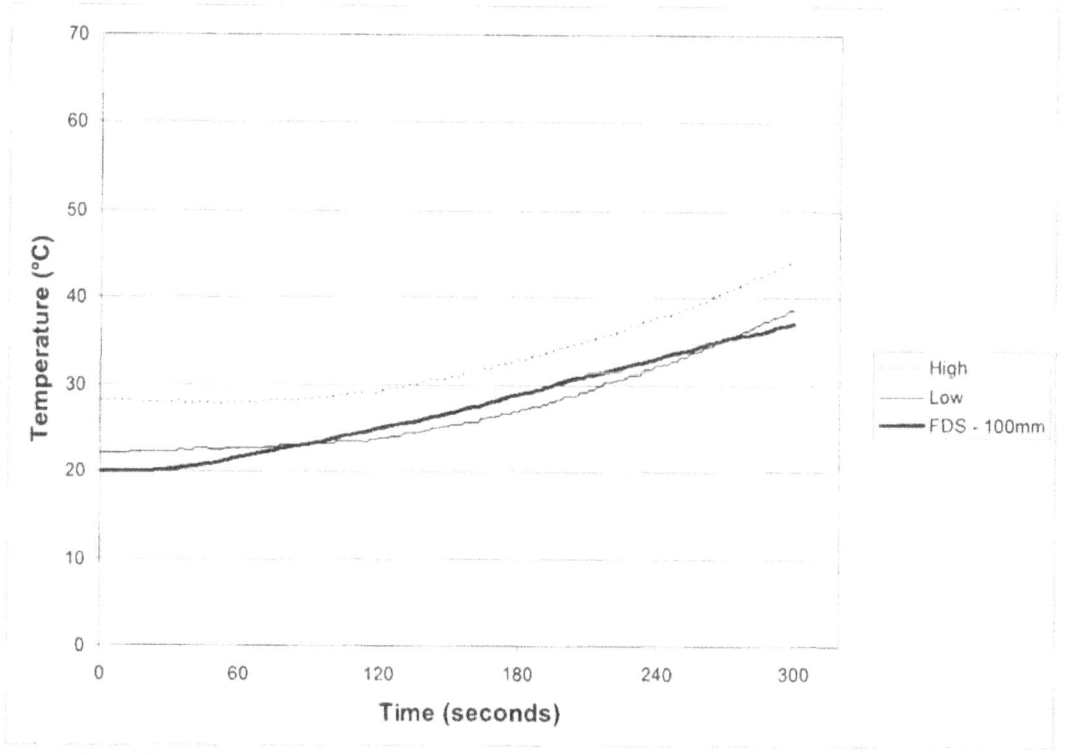

Figure B.92 – Comparison of Predicted and Measured Heat Detector Temperatures, RTI = 287 m$^{1/2}$-s$^{1/2}$, 10.7 m Ceiling Height, Radial Distance = 6.5 m

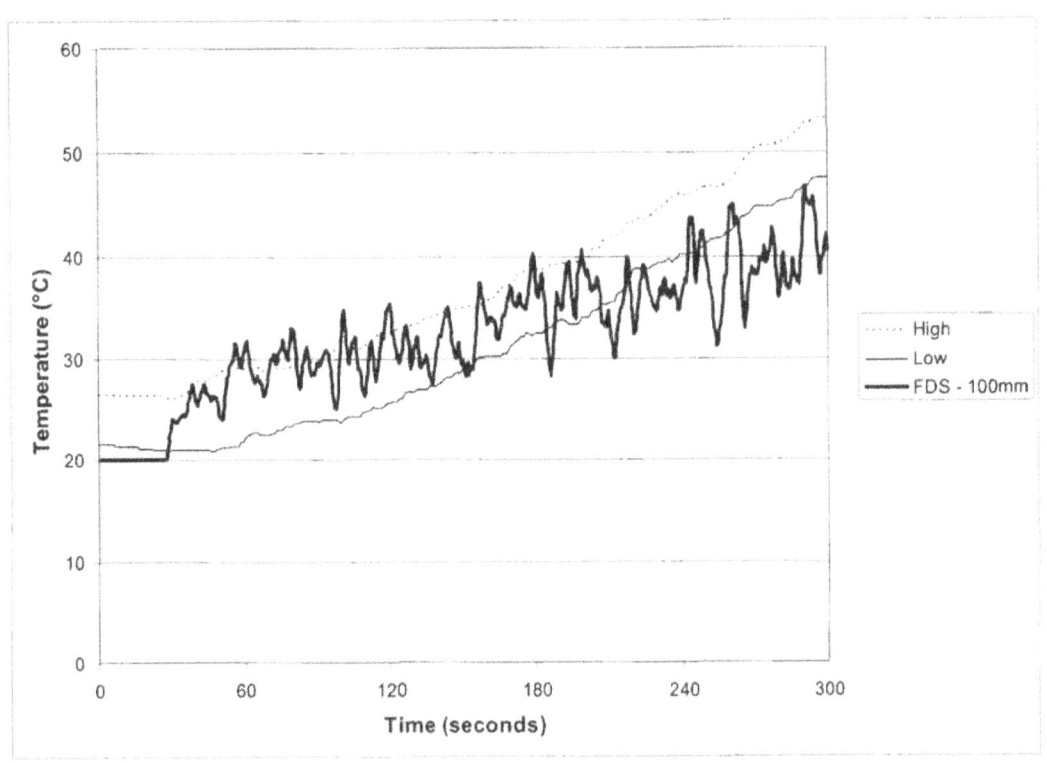

Figure B.93 – Comparison of Predicted and Measured Temperatures, 10.7 m Ceiling Height, Radial Distance = 10.8 m

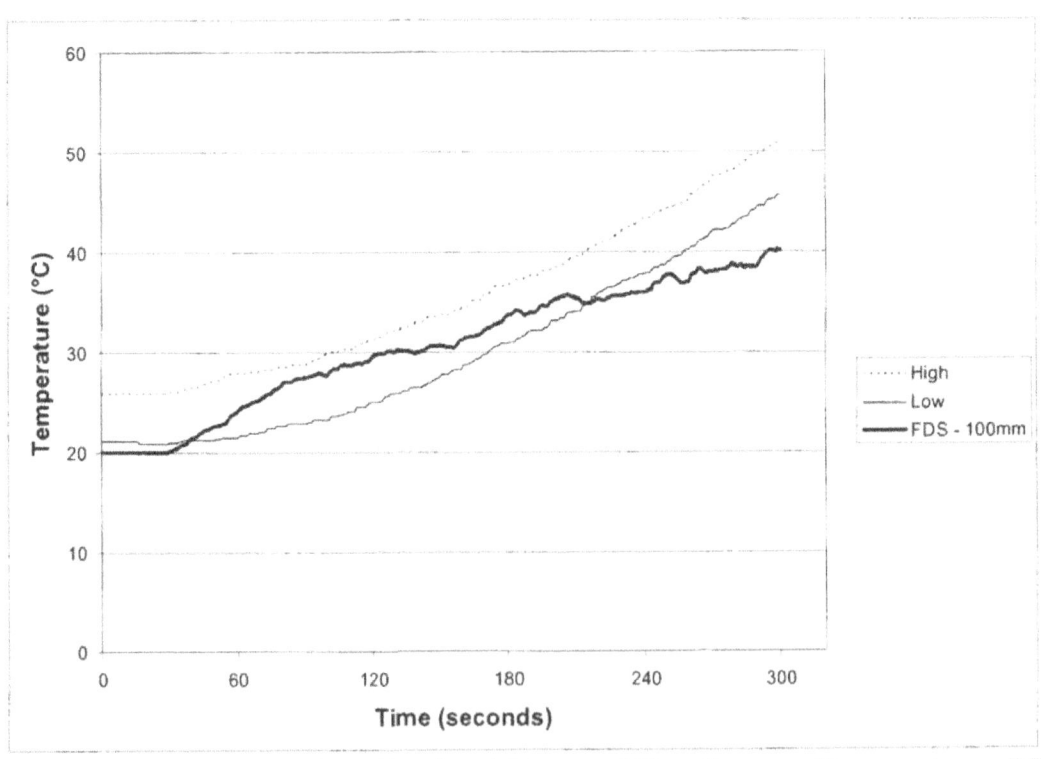

Figure B.94 – Comparison of Predicted and Measured Heat Detector Temperatures, RTI = 32 m$^{1/2}$-s$^{1/2}$, 10.7 m Ceiling Height, Radial Distance = 10.8 m

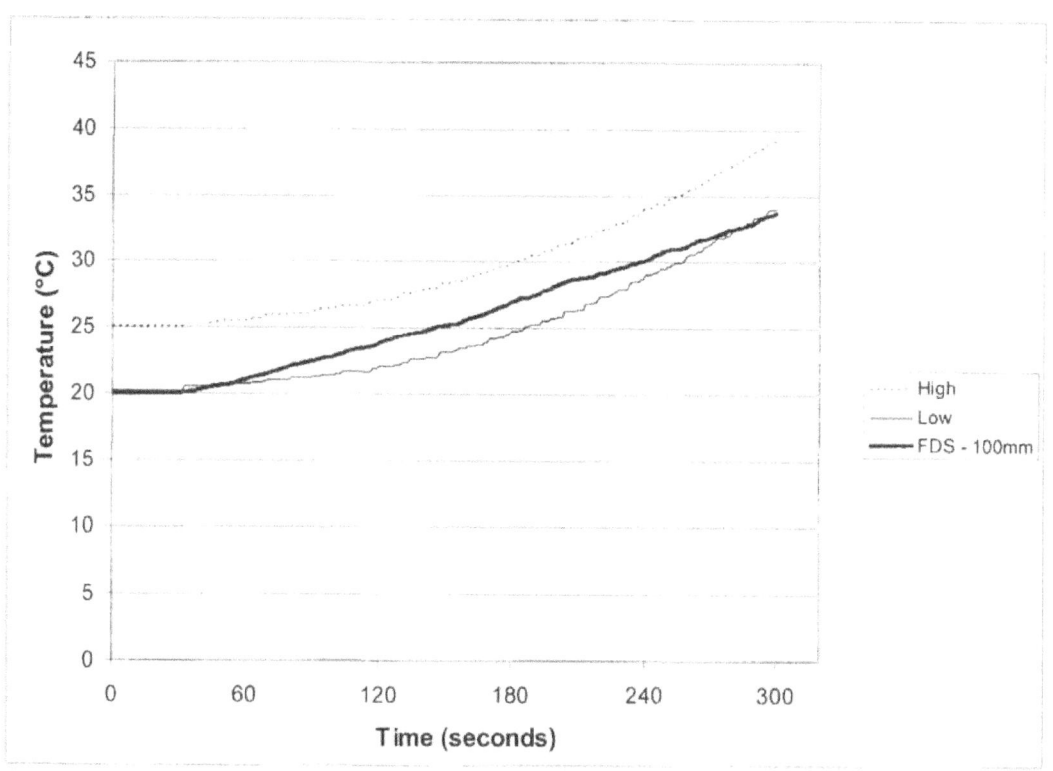

Figure B.95 – Comparison of Predicted and Measured Heat Detector Temperatures, RTI = 164 m$^{1/2}$-s$^{1/2}$, 10.7 m Ceiling Height, Radial Distance = 10.8 m

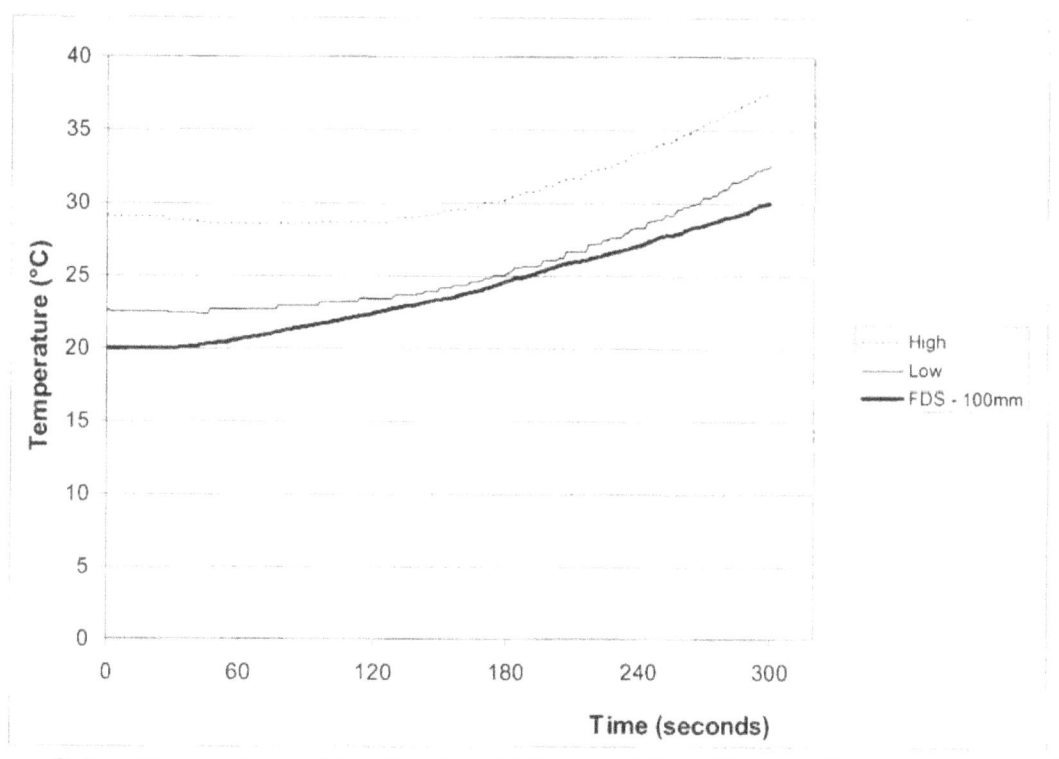

Figure B.96 – Comparison of Predicted and Measured Heat Detector Temperatures, RTI = 287 m$^{1/2}$-s$^{1/2}$, 10.7 m Ceiling Height, Radial Distance = 10.8 m

www.ingramcontent.com/pod-product-compliance
Lightning Source LLC
Chambersburg PA
CBHW080306180526
45167CB00006B/2693